Ana Irene Ramírez Galarza
José Seade

Introduction to Classical Geometries

Birkhäuser
Basel · Boston · Berlin

Authors:

Ana Irene Ramírez Galarza
Departamento de Matemáticas
Facultad de Ciencias
Universidad Nacional Autónoma de
México
Ciudad Universitaria, Circuito Exterior
México 04510 D. F.
México
e-mail: anai@matematicas.unam.mx

José Seade
Instituto de Matemáticas
Unidad Cuernavaca
Universidad Nacional Autónoma de México
Av. Universidad s/n
Ciudad Universitaria,
Lomas de Chamilpa
Cuernavaca, Morelos
México
e-mail: jseade@matem.unam.mx

Translated from the Spanish original "Introducción a la Geometría Avanzada"
© 2002 Las Prensas de Ciencia, UNAM, México

Figures by Juan Pablo Romero Méndez

2000 Mathematical Subject Classification: 51, 53, 14, 22, 83

Library of Congress Control Number: 2007922251

Bibliographic information published by Die Deutsche Bibliothek
Die Deutsche Bibliothek lists this publication in the Deutsche Nationalbibliografie; detailed bibliographic data is available in the Internet at http://dnb.ddb.de

ISBN 978-3-7643-7517-1 Birkhäuser Verlag AG, Basel – Boston – Berlin

This work is subject to copyright. All rights are reserved, whether the whole or part of the material is concerned, specifically the rights of translation, reprinting, re-use of illustrations, recitation, broadcasting, reproduction on microfilms or in other ways, and storage in data banks. For any kind of use permission of the copyright owner must be obtained.

© 2007 Birkhäuser Verlag AG
Basel • Boston • Berlin
P.O. Box 133, CH-4010 Basel, Switzerland
Part of Springer Science+Business Media
Printed on acid-free paper produced from chlorine-free pulp. TCF∞
Printed in Germany
ISBN-10: 3-7643-7517-5   e-ISBN-10: 3-7643-7518-3
ISBN-13: 978-3-7643-7517-1   e-ISBN-13: 978-3-7643-7518-8

9 8 7 6 5 4 3 2 1   www.birkhauser.ch

Contents

Preface   vii

List of Symbols   x

1 Euclidean geometry   1
    1.1 Symmetries 2
    1.2 Rigid transformations 15
    1.3 Invariants under rigid transformations 28
    1.4 Cylinders and tori 37
    1.5 Finite subgroups of $E(2)$ and $E(3)$ 46
    1.6 Frieze patterns and tessellations 58

2 Affine geometry   75
    2.1 The line at infinity 76
    2.2 Affine transformations and their invariants 83

3 Projective geometry   91
    3.1 The real projective plane 92
    3.2 The Duality Principle 99
    3.3 The shape of $P^2(\mathbb{R})$ 103
    3.4 Coordinate charts for $P^2(\mathbb{R})$ (and for $P^1(\mathbb{C})$) 109
    3.5 The projective group 113
    3.6 Invariance of the cross ratio 121
    3.7 The space of conics 126
    3.8 Projective properties of the conics 129
    3.9 Poles and polars 134
    3.10 Elliptic geometry 141

4 Hyperbolic geometry   149
    4.1 Models of the hyperbolic plane 149
    4.2 Transformations of the hyperbolic plane 157
    4.3 Steiner network 164

	4.4 The hyperbolic metric	168
	4.5 First results in hyperbolic geometry	177
	4.6 Surfaces with hyperbolic structure	183
	4.7 Tessellations	192
5	**Appendices**	**199**
	5.1 Differentiable functions	199
	5.2 Equivalence relations	201
	5.3 The symmetric group in four symbols: S_4	202
	5.4 Euclidean postulates	205
	5.5 Topology	206
	5.6 Some results on the circle	208
Bibliography		**211**
Index		**215**

Preface

Geometry is one of the oldest branches of mathematics, nearly as old as human culture. Its beauty has always fascinated mathematicians, among others. In writing this book we had the purpose of sharing with readers the pleasure derived from studying geometry, as well as giving a taste of its importance, its deep connections with other branches of mathematics and the highly diverse viewpoints that may be taken by someone entering this field.

We also want to propose a specific way to introduce concepts that have arisen from the heyday of the Greek school of geometry to the present day. We work with coordinate models, since this facilitates the use of algebraic and analytic results, and we follow the viewpoint proposed by Felix Klein in the 19th century, of studying geometry via groups of symmetries of the space in question.

We intend this book to be both an introduction to the subject addressed to undergraduate students in mathematics and physics, and a useful text-book for mathematicians and scientists in general who want to learn the basics of classical geometry: Euclidean, affine, elliptic, hyperbolic and projective geometry. These are all presented in a unified way and the essential content of this book may be covered in a single semester, though a longer period of study would allow the student to grasp and assimilate better the material in it.

We essentially restrict the whole discussion to "plane geometry", that is dimension 2, since this is already rich enough. We include some aspects of the 3- or n-dimensional extension of these plane geometries whenever it is simple to do so. Once plane geometry is well understood, it is much easier to go into higher dimensions, and we give guidelines for further reading.

We assume that the reader is familiar with a few facts of plane Euclidean geometry (especially those concerning triangles and circles), and also with the elements of analytic geometry (the intersection of straight lines and planes, their equations and the main properties of conics and quadrics) and of linear algebra (up to eigenvalues and eigenvectors of a linear transformation of the plane or 3-space).

Also, we suppose the student is familiar with the notions and results of differential and integral calculus for real functions of a single real variable, and that he or she is taking at least a first course in calculus of several variables; basic knowledge of the complex numbers will be useful in the last chapter.

With this background it is possible to present formally, applying the analytic method to coordinate models, the geometries that "come after" Euclidean geometry: affine, projective, elliptic and hyperbolic.

The study we make of these geometries allows us to understand the fundamental role played by the groups of "symmetries" and shows how algebra, analysis and geometry combine to give a better understanding of a concept, a result or a problem's solution.

We have also paid attention to the exercises, choosing them so that their solution deepens or widens our understanding. For exercises of greater than average difficulty we give references as support. The best measure of the degree of understanding is the percentage of exercises solved: all of them should be attempted!

After assimilating the material in this text the reader will be prepared to step forward, if that is his or her will, into more advanced areas of geometry, such as differential geometry, foliations and group actions on differentiable manifolds, among others.

Dedication

*In memory of Sevín Recillas Pishmish; and to Francisco González Acuña, Santiago López de Medrano and Alberto Verjovsky. The four of them taught us the beauty of geometry and they made a significant contribution to **turning** geometry into a living science in México.*

Acknowledgments

We are grateful to Juan Pablo Romero Méndez for the excellent illustrations, to Ricardo Berlanga, Laura Ortiz Bobadilla, Ernesto Rosales González, Oscar Palmas Velasco, León Kushner Schnur and Alberto Verjovsky for many useful conversations and comments, and to David Mireles Morales, Rolando Gómez Macedo, Ernesto Mayorga Saucedo, Christian Rubio Montiel, Juan Carlos Fernández Morales and Jesús Núñez Zimbrón, for their help and support.

List of Symbols

$\mathbb{N}$	natural numbers
$\mathbb{Z}$	integer numbers
$\mathbb{Q}$	rational numbers
$\mathbb{R}$	real numbers
$\mathbb{C}$	complex numbers
$\mathbb{R}^n$	n-dimensional Cartesian space
S^n	n-sphere: vectors of norm 1 in $\mathbb{R}^{n+1}$
S_n	symmetric group in n symbols
Ro_θ	rotation by an angle θ in $\mathbb{R}^2$ about the origin
Re_ϕ	reflection with respect to the straight line through the origin with slope $\tan\phi$ in $\mathbb{R}^2$
$E(n)$	group of rigid transformations in $\mathbb{R}^n$
$GL(n, \mathbb{R})$	linear group of order n
$SL(n, \mathbb{R})$	special linear group of order n
$O(n, \mathbb{R})$	orthogonal group in $\mathbb{R}^n$
$SO(n, \mathbb{R})$	group of rotations about the origin in $\mathbb{R}^n$
A^2	affine plane
$A(2)$	affine group
$P^n(\mathbb{R})$	n-dimensional projective space on $\mathbb{R}$
$P^1(\mathbb{C})$	one-dimensional projective space on $\mathbb{C}$
D^n	vectors of norm less than 1 in $\mathbb{R}^n$
$\nabla F(P_0)$	gradient of F in P_0
$PGL(n, \mathbb{R})$	projectivization of the linear group of order n
$PSL(2, \mathbb{C})$	group of Möbius transformations
Δ	model of the disk for hyperbolic geometry
H^+	model of the upper half plane for hyperbolic geometry
$G_\mathcal{C}$	subgroup of $PGL(3, \mathbb{R})$ that leaves a conic fixed
G_Δ^+	subgroup of $PSL(2, \mathbb{C})$ that leaves Δ fixed
$G_{H^+}^+$	$PSL(2, \mathbb{R})$, subgroup of $PSL(2, \mathbb{C})$ that leaves H^+ fixed
G_Δ	isometries of Δ
G_{H^+}	isometries of H^+

Euclidean geometry

The branch of mathematics known as geometry was formally established in Greece around the year 300 B.C. However its origins, for our Western culture, go back to Mesopotamia and Egypt around the year 3000 B.C.

Euclid's classical treatise "The Elements" [Eu] has paramount importance for the whole of science because in addition to being a compilation and organization of all geometrical and physical knowledge generated up to that time, it also presents a method of validating theoretical knowledge, making it imperishable. This is why that treatise is still edited in our own days.

That text had small flaws which took a long time to be corrected (see [H]). However, it is amazing to realize that most of those flaws disturbed Euclid himself, who, at each occasion, handled the situation with all the caution the culture in which he was embedded allowed him. In Euclid's culture, fundamental concepts such as real number, limit, and group of transformations did not yet exist. Today we can use these concepts because of the algebraic language we have available.

We do not discuss the Euclidean postulates here (see Appendix 5.4), but focus on the analytic treatment of geometry which assigns coordinates to points, equations to loci, and in which the permitted transformations are conceived as functions. That means that we shall base our study of geometry on the properties of the system of real numbers that are studied in a first college course of calculus (see [Cou]). With those properties, it is possible to prove that the Cartesian plane complies with the Euclidean postulates (Exercise 17).

According to various authors (see, for instance, [Ki]), Euclid felt reluctant about the *method of superposition* he used in his proofs of congruences of triangles. Nevertheless, the analytic method clarifies Euclid's "principle of superposition" up to making it the very essence of Euclidean geometry: the study of invariants under rigid transformations.

This first chapter is devoted to the study of the group of rigid transformations in $\mathbb{R}^2$ and $\mathbb{R}^3$, and presents a view of geometric results that can be obtained under the analytic approach. The main references are [Cox 1, 2, 5], [Eu], [Ev], [H], [Mar], [Ra], and [Re].

1.1 Symmetries

The concept of symmetry is fundamental in geometry as well as in nature itself. The human body is outwardly symmetric with respect to a plane, and that symmetry determined the construction of objects which are symmetric as well, such as vases with two handles or pairs of shoes. The symmetry of a disk with respect to its center provides many applications. An infinite circular cylinder is symmetric not just with respect to many planes and many points, but also with respect to many straight lines; in particular, with respect to its axis. In this section, we shall show how to justify these last assertions on the basis of the equation of a cylinder.

To be precise about what should be understood by each type of symmetry, we start by recalling how we determine some distances in three-dimensional space. We recall those formulas further on.

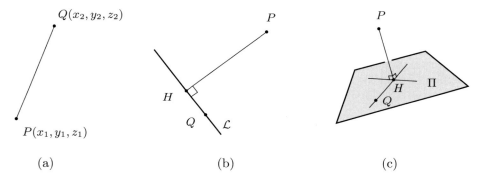

Figure 1.1: Distances: (a) from a point P to a point Q, (b) from a point P to a straight line $\mathcal{L}$, and (c) from a point P to a plane Π

Definition. The **distance from a point** P **to a point** Q is the length of the rectilinear segment between the two points.

Definition. The distance from a point P to a straight line L is the length of the perpendicular segment from the point to the line.

Note that from all the points in the straight line, the foot H of the perpendicular from point P to the straight line $\mathcal{L}$ determines a rectilinear segment of minimal length (it is an edge of any right triangle PHQ in Figure 1.1(b)).

Definition. The distance from a point P to a plane p is the length of the perpendicular rectilinear segment from the point to the plane.

In this case, the foot H of the perpendicular from point P to the plane Π is also the point of the plane which minimizes the length of all possible rectilinear segments from P to a point Q of the plane, since PH is an edge of any of the right triangles PHQ in Figure 1.1(c).

1.1. Symmetries

The different types of symmetry an object in the 3-dimensional space can have are the following: (a) with respect to a point, (b) with respect to a straight line, and (c) with respect to a plane.

Definition. An object $\mathcal{F}$ is **symmetric with respect to a point** O if for each point P in $\mathcal{F}$, P' is also in $\mathcal{F}$, where O is the midpoint of the rectilinear segment PP' (see Figure 1.2). The point O is called the **center of symmetry**.

The graph of the sine function is symmetric with respect to the origin (see Figure 1.2(a)). A cone of revolution is symmetric with respect to its vertex (see Figure 1.2(b)). A cube is symmetric with respect to its center (see Figure 1.2(c)). The analytic proof of the first two assertions is very simple, as will be seen.

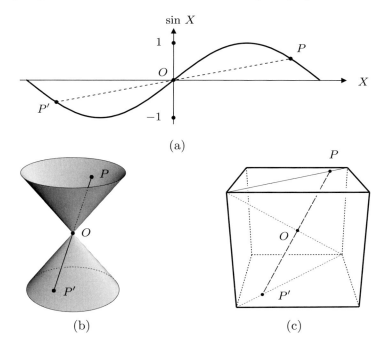

Figure 1.2: Figures symmetric with respect to a point.

Definition. An object $\mathcal{F}$ is **symmetric with respect to a straight line** $\mathcal{L}$ if for each point P in $\mathcal{F}$, P' is also in $\mathcal{F}$, where $\mathcal{L}$ intersects perpendicularly the rectilinear segment PP' (see Figure 1.3). The straight line $\mathcal{L}$ is an **axis of symmetry**.

The graph of the cosine function is symmetric with respect to the Y axis (see Figure 1.3(a)), but not with respect to the X axis. A regular pentagon is symmetric with respect to any straight line through one vertex and the midpoint of the side opposite to that vertex (as in Figure 1.3(b)). A cube is symmetric with respect to each of the straight lines through the centers of opposite faces (see Figure 1.3(c)). At the end of this section, we shall prove these assertions very easily.

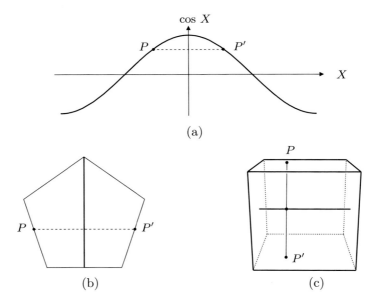

Figure 1.3: Figures symmetric with respect to a straight line.

Definition. An object $\mathcal{F}$ is **symmetric with respect to a plane** Π if for each point P in $\mathcal{F}$, P' is also in $\mathcal{F}$, where Π is the plane perpendicular to the rectilinear segment PP' through its midpoint (see Figure 1.4). The plane Π is a **plane of symmetry**.

The human body is outwardly symmetric with respect to the plane through the spinal column and the tip of the nose. A cube is symmetric with respect to a plane containing diagonals parallel to opposite faces, and with respect to a plane parallel to opposite faces and through the center. An infinite circular cylinder is symmetric with respect to any plane perpendicular to its axis, and with respect to any plane through its axis. We shall verify the latter two assertions at the end of this section.

There are formulas to compute the distances involved in the definitions of symmetry. Nevertheless, let us see, before revising them, how some simple algebraic considerations suffice to obtain the point symmetric to a given point $P(x, y, z)$ with respect to a coordinate plane, a coordinate axis, and to the origin.

This is useful because — except in those cases when the coordinate system is given in advance — we shall be able to take the coordinate system so that the plane, the straight line or the point with respect to which we are interested in examining the symmetry, will be one of those coordinate elements. Furthermore,

1.1. Symmetries 5

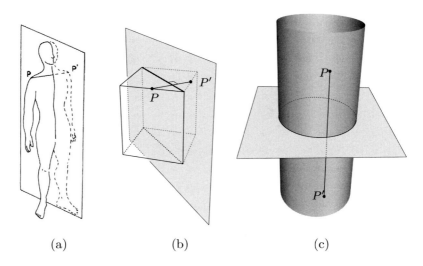

(a)   (b)   (c)

Figure 1.4: Figures symmetric with respect to a plane.

the standard inclusion of $\mathbb{R}^2$ in $\mathbb{R}^3$ allows us to use those criteria in the case of the plane. The following definitions are illustrated in Figure 1.5.

The symmetric point of $P(x, y, z)$ **with respect to the origin of coordinates** O, is the point $P_O(-x, -y, -z)$, since P, P_O and O are collinear, $-P = P_O$ and $||OP|| = ||OP_O||$.

The symmetric point of $P(x, y, z)$ **with respect to the** X **axis**, is the point $P_X(x, -y, -z)$, since the vector $P - P_X = (0, 2y, 2z)$ is perpendicular to $(1, 0, 0)$ and the midpoint of the rectilinear segment PP_X is $(x, 0, 0) \in X$.

Analogously, it is proved that **the point symmetric to** $P(x, y, z)$ **with respect to the** Y **axis** is the point $P_Y(-x, y, -z)$, and that **the point symmetric to** $P(x, y, z)$ **with respect to the** Z **axis** is the point $P_Z(-x, -y, z)$.

It is also easy to prove that **the points symmetric to** $P(x, y, z)$ with respect to each of the coordinate planes are as follows:

$P_{XY}(x, y, -z)$ is symmetric to $P(x, y, z)$ with respect to the XY plane,

$P_{YZ}(-x, y, z)$ is symmetric to $P(x, y, z)$ with respect to the YZ plane,

$P_{ZX}(x, -y, z)$ is symmetric to $P(x, y, z)$ with respect to the ZX plane.

On the basis of the preceding paragraphs, the reader should not find any difficulty in proving the following result:

If a given figure is symmetric with respect to the three coordinate planes, then it is also symmetric with respect to the coordinate axes and to the origin.

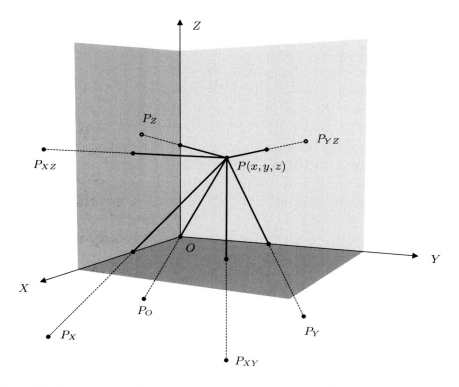

Figure 1.5: Points symmetric to a given point with respect to the coordinate planes and axes, and to the origin.

Now let us study the necessary formulas and algebra to determine the symmetries we have mentioned above.

For any two given vectors in $\mathbb{R}^3$ we define their **scalar product** as

$$(x_1, y_1, z_1) \cdot (x_2, y_2, z_2) = x_1 x_2 + y_1 y_2 + z_1 z_2.$$

By means of this scalar product, we obtain the **norm of a vector** (x, y, z) in $\mathbb{R}^3$:

$$\|(x, y, z)\| = \sqrt{(x, y, z) \cdot (x, y, z)} = \sqrt{x^2 + y^2 + z^2},$$

which can be interpreted as the length of the diagonal of the parallelepiped determined by $(0, 0, 0)$, (x, y, z), and the projections of (x, y, z) on each of the coordinate planes and on each of the coordinate axes (make a drawing).

By means of the norm we can define **the distance between two points** $P(x_1, y_1, z_1)$ and $Q(x_2, y_2, z_2)$ as:

$$d(P, Q) = \|P - Q\| = \sqrt{(x_1 - x_2)^2 + (y_1 - y_2)^2 + (z_1 - z_2)^2},$$

1.1. Symmetries

and so the **angle between two vectors** $\bar{u} = (u_1, u_2, u_3)$ and $\bar{v} = (v_1, v_2, v_3)$ can be defined as:

$$\angle(\bar{u}, \bar{v}) = \angle \cos\left(\frac{\bar{u} \cdot \bar{v}}{||\bar{u}||\,||\bar{v}||}\right),$$

since **Schwarz's inequality** asserts that (see Exercise 7)

$$|\bar{u} \cdot \bar{v}| \leq ||\bar{u}||\,||\bar{v}||.$$

Remark 1. In this book we identify $\mathbb{R}^2$ with the image of $\mathbb{R}^2$ in $\mathbb{R}^3$ given by $(x, y) \mapsto (x, y, 0)$. Of course, this is not the only way of considering $\mathbb{R}^2$ as a vector subspace of $\mathbb{R}^3$, but this is the most commonly used.

In addition to the scalar product of two vectors $\bar{u} = (u_1, u_2, u_3)$ and $\bar{v} = (v_1, v_2, v_3)$ in $\mathbb{R}^3$, we also have their **cross product**

$$\bar{u} \times \bar{v} = (u_2 v_3 - u_3 v_2, u_3 v_1 - u_1 v_3, u_1 v_2 - u_2 v_1) = \begin{vmatrix} \hat{\imath} & \hat{\jmath} & \hat{k} \\ u_1 & u_2 & u_3 \\ v_1 & v_2 & v_3 \end{vmatrix},$$

where $\hat{\imath}$, $\hat{\jmath}$ and $\hat{k}$ are the vectors of the **standard base** of $\mathbb{R}^3$: $(1, 0, 0)$, $(0, 1, 0)$ and $(0, 0, 1)$, respectively.

By means of the norm of the cross product, we can compute the **area of a parallelogram**: if $\theta = \angle(\bar{u}, \bar{v})$, then we easily prove that

$$||\bar{u} \times \bar{v}|| = ||\bar{u}||\,||\bar{v}||\,|\sin \theta|.$$

Finally, the **scalar triple product** of three vectors $\bar{u}, \bar{v}, \bar{w}$, denoted as $[\bar{u}, \bar{v}, \bar{w}]$, is used for computing the **oriented volume of the parallelepiped** determined by $\bar{0}$, $\bar{u}, \bar{v}, \bar{w}, \bar{u} + \bar{v}, \bar{u} + \bar{w}, \bar{v} + \bar{w}$ and $\bar{u} + \bar{v} + \bar{w}$ (see Figure 1.6):

$$[\bar{u}, \bar{v}, \bar{w}] = \bar{u} \cdot \bar{v} \times \bar{w} = ||\bar{u}||\,||\bar{v} \times \bar{w}||\cos \phi,$$

where $\phi = \angle(\bar{u}, \bar{v} \times \bar{w})$. Since $\bar{v} \times \bar{w}$ is perpendicular to $\bar{v}$ and $\bar{w}$, the number $||\bar{u}|| \cos \phi$ is the oriented height of $\mathcal{P}(\bar{u}, \bar{v}, \bar{w})$ with respect to the basis formed by the parallelogram determined by $\bar{v}$ and $\bar{w}$ (see Figure 1.6), whose area is precisely $||\bar{v} \times \bar{w}||$.

Now it is easy to establish a formula for computing the **distance from a given point** $Q(a, b, c)$ to a line $\mathcal{L} \subset \mathbb{R}^3$ through a point P_0 in the direction of a unit vector $\hat{u}$:

$$d(Q, \mathcal{L}) = ||(Q - P_0) \times \hat{u}|| = ||Q - P_0||\,|\sin \theta| \tag{1.1}$$

since $||\hat{u}|| = 1$, the norm of $(Q - P_0) \times \hat{u}$ is equal to $||Q - P_0||\,|\sin \theta|$ which is precisely the height of the parallelogram shown in Figure 1.7, and which is contained in the plane defined by $\mathcal{L}$ and Q, where $\theta = \angle(\bar{u}, Q - P_0)$.

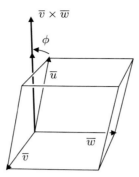

Figure 1.6: Geometric interpretation of the scalar triple product.

If the straight line and the point are in $\mathbb{R}^2$, the formula for **the distance from the point $Q(x_0, y_0)$ to a straight line $\mathcal{L} \subset \mathbb{R}^2$**, represented by the equation $Ax + By + C = 0$, is

$$d(Q, \mathcal{L}) = \frac{|Ax_0 + By_0 + C|}{\sqrt{A^2 + B^2}}. \tag{1.2}$$

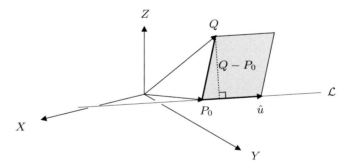

Figure 1.7: How to compute the distance from a point to a straight line in $\mathbb{R}^3$.

The reader ought to justify this formula in terms of the scalar product and, using the same reasoning, prove that the formula for the distance from a point $Q(x_0, y_0, z_0)$ to a plane Π, represented by the equation $Ax + By + Cz + D = 0$, is

$$d(Q, \Pi) = \frac{|Ax_0 + By_0 + Cz_0 + D|}{\sqrt{A^2 + B^2 + C^2}}. \tag{1.3}$$

Since we already have a way of determining distances between points, angles between straight lines, areas, and volumes, we are ready to begin with the analytic study of Euclidean geometry in the plane and in three-dimensional space.

1.1. Symmetries

We are going to prove first that we have the symmetries described in all the previous examples, selecting in each case an **adequate coordinate system** so that we can apply the criteria we have just recalled. There is no general recipe to determine how to make this choice; in this situation, as in many others, it is only through practice that we develop an intuition that allows us to make these choices appropriately.

1. The graph of the sine function is symmetric with respect to the origin.

In this case, the coordinate system has already been established, since the graph of a function $y = f(x)$ refers to the coordinates (x, y) of a coordinate plane.

The graph of the sine function is formed by the points $(x, \sin x)$, and since $\sin x = -\sin(-x)$, we can write $-\sin x = \sin(-x)$. Thus, a point $P(x, \sin x)$ belongs to the graph of the sine function if and only if the point $P'(-x, -\sin x)$ also belongs to that graph (see Figure 1.8).

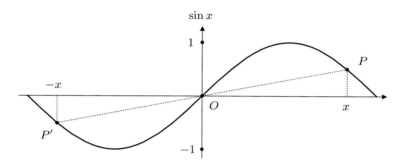

Figure 1.8: The graph of the sine function is symmetric with respect to the origin.

This property of the sine function reminds us of the property of the functions $y = x^n$, where n is an odd integer; hence it is said that the sine function is an odd function.

2. The graph of the cosine function is symmetric with respect to the Y axis.

In this case, the coordinate system is also given. We invite the reader to sketch the graph and to recall the trigonometric identity which implies that a point $P(x, \cos x)$ belongs to the graph of the cosine function if and only if the point $P'(-x, \cos(-x))$ also belongs to that graph.

3. Every circle is symmetric with respect to its center and with respect to each of its diameters.

In this case, it is convenient to take as a coordinate system one whose origin is the center and whose X axis *coincides with the diameter with respect to which we want to prove the symmetry*. Thus, the equation of the circle is

$$x^2 + y^2 = r^2$$

and since $x^2 = (-x)^2$ and $y^2 = (-y)^2$, it is obvious that $P(x,y)$ satisfies the equation of the circle if and only if the point $P'(-x,-y)$ also satisfies it; that is, the circle is symmetric with respect to its center. For the same reason, $P_X(x,-y)$ belongs to the circle if and only if $P(x,y)$ also belongs to the circle. This proves the symmetry of the circle with respect to any of its diameters (see Figure 1.9).

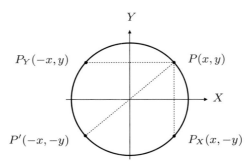

Figure 1.9: Symmetries of the circle.

4. A regular pentagon is symmetric with respect to the straight lines through one of its vertices and the midpoint of the side opposite to that vertex.

Recall that a pentagon is regular if its sides are congruent to one another and its interior angles are congruent to one another.

To prove this claim, first, we can take a coordinate system suchthat if A, B, C, D and E are the vertices of the pentagon, A is chosen on the Y axis and the vertices C and D symmetric with respect to the Y axis. Thus, the axis is one of the straight lines we are interested in (see Figure 1.10).

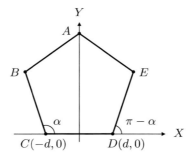

Figure 1.10: Symmetries of the pentagon.

Because of the congruence of the interior angles at C and D, the sides BC and ED form angles $\pi - \alpha$ and α, respectively, with the positive X axis.

1.1. Symmetries

Therefore, if the slope of DE is m, then the slope of BC is $-m$, and since they pass through the points C and D respectively, the equations of the sides BC and DE are
$$BC: y = -mx - md, \quad DE: y = mx - md\,.$$

With these equations, it is easy to prove that the points of the side CB have their symmetric points with respect to the Y axis on the side DE. We leave the proof of this property to the reader. It is also left to the reader to prove that the symmetric points of the side BA are on the side EA.

5. A circular cylinder is symmetric with respect to its axis, to any plane through its axis, to any straight line and any plane which cuts the axis perpendicularly, and with respect to any point on the axis.

We can think of a circular cylinder as the surface of revolution generated by a straight line rotating about a fixed parallel straight line, the axis of revolution.

In this case, we place the cylinder in such a way that the Z axis is its axis; the plane through the Z axis with respect to which we want to prove the symmetry is the YZ plane; the plane orthogonal to the axis of the cylinder, with respect to which we want to examine the symmetry, is the XY plane, and the proposed center of symmetry is the origin.

With all these specifications, the equation of the cylinder is
$$x^2 + y^2 = r^2;$$
since x and y have an even exponent and z is free, it is clear that there exists symmetry with respect to the origin, to each of the coordinate axes, and to each of the coordinate planes, in particular, with respect to the Z axis, to the YZ plane, and to the XY plane.

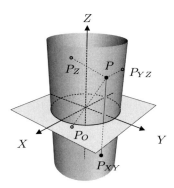

Figure 1.11: Symmetries of a cylinder of revolution.

At the beginning, the reader might not have believed that there are axes of symmetry other than the axis of revolution. However, the simplicity of the criterion for proving the symmetry of a figure with respect to a coordinate axis led us to discover that the cylinder has an infinite number of axes of symmetry: each straight line that intersects the axis of revolution perpendicularly.

That is one of the advantages of studying geometry using algebra: this allows us to discover geometric facts that we might not have imagined or believed before.

6. A cone of revolution — which is generated by complete straight lines — is symmetric with respect to its vertex, to its axis, and with respect to each straight line through its vertex and perpendicular to its axis. It is also symmetric with respect to the plane through its vertex and perpendicular to its axis, and to each plane containing the axis.

In this case, it is also convenient to take the Z axis as the axis of revolution, the origin as the vertex, the plane perpendicular to the axis with respect to which we are interested in analyzing the symmetry as the XY plane, and the plane through the axis of revolution as the YZ plane.

The reader must make a drawing, establish the equation of the cone, and verify that every assertion of the proposition is satisfied because the cone of revolution turns out to be symmetric with respect to all the coordinate elements.

7. A cube is symmetric with respect to the planes equidistant from two parallel faces, to the planes containing parallel diagonals of parallel faces, to the straight lines through the centers of opposite faces, to the straight lines through the midpoints of opposite edges, and to the point where all planes and axes of symmetry meet.

Let us consider a cube whose edges have length $2a$. Let us place the cube in such a way that the origin is the point where the equidistant planes of opposite faces meet (see Figure 1.12), and let us take as coordinate planes precisely those equidistant planes.

The equations of the planes containing the faces are $x = a$ and $x = -a$, $y = a$ and $y = -a$, $z = a$ and $z = -a$; furthermore, if the vertices of the top are A, B, C and D, and those of the base are R, S, T, and U, then the faces are defined as follows:

$ABCD = \{(x,y,z)|\, z = a,\, |x| \leq a,\, |y| \leq a\};$

$RSTU = \{(x,y,z)|\, z = -a,\, |x| \leq a,\, |y| \leq a\};$

$BSTC = \{(x,y,z)|\, y = a,\, |x| \leq a,\, |z| \leq a\};$

$ARUD = \{(x,y,z)|\, y = -a,\, |x| \leq a,\, |z| \leq a\};$

$ARSB = \{(x,y,z)|\, x = a,\, |y| \leq a,\, |z| \leq a\};$

$DUTC = \{(x,y,z)|\, x = -a,\, |y| \leq a,\, |z| \leq a\}.$

The coordinates of a point P on the face $ABCD$ are (x, y, a), where $|x| \leq a$ and $|y| \leq a$; thus, its symmetric point with respect to the plane XY is $P_{XY}(x, y, -a)$, which belongs to the face $RSTU$.

1.1. Symmetries

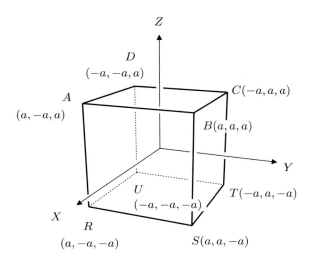

Figure 1.12: Symmetries of a cube.

The symmetries with respect to the YZ plane, which is equidistant to the front and back faces, and to the ZX plane, which is equidistant to the right and left faces, are proved analogously.

Thus, according to the remark following the list of criteria for symmetry with respect to the coordinate elements, we have also proved the symmetries with respect to the coordinate axes — the straight lines joining the centers of opposite faces, and to the origin — the meeting point of the three straight lines joining the centers of opposite faces intersect. This point is called the **center of the cube** because it is a center of symmetry.

To prove that the cube is symmetric with respect to the planes determined by parallel diagonals of parallel faces, it suffices to characterize, in an appropriate way, the points belonging to the faces which are at both sides of one of those planes. The behavior of the parameters will then facilitate verifying the symmetry.

Let us consider as an example the plane $ARTC$ (see Figure 1.12), which clearly has $(1, 1, 0)$ as a normal vector and, since it passes through the origin, satisfies the equation $x + y = 0$.

Figure 1.12 shows that the face $ARUD$ is symmetric to the face $ARSB$ with respect to the plane $ARTC$.

The points of the plane $ARUD$ are of the form $P = A + s\widehat{u} + t\widehat{v}$, where $\widehat{u}$ is a unit vector in the direction of $A - D = (2a, 0, 0)$ and $\widehat{v}$ is a unit vector in the direction of $A - R = (0, 0, 2a)$:

$$(x, y, z) = (a, -a, a) + s(1, 0, 0) + t(0, 0, 1) = (a + s, -a, a + t).$$

The points are confined to the face $ARUD$ when $-2a \leq s \leq 0$ and $-2a \leq t \leq 0$.

By a similar argument it is shown that the points P' on the face $ARSB$ are of the form
$$(x, y, z) = (a, -a, a) + \sigma(0, 1, 0) + \tau(0, 0, 1) = (a, -a + \sigma, a + \tau),$$
where $0 \leq \sigma \leq 2a$ and $-2a \leq \tau \leq 0$.

When $s = -\sigma$ and $t = \tau$, the segment PP' is perpendicular to the plane $ARTC$, since $P - P' = (s, s, 0)$ is parallel to $(1, 1, 0)$. By the formula (1.3), the distances from P and P' to $ARTC$ are, respectively,
$$\frac{|a - s - a|}{\sqrt{2}} = \frac{|s|}{\sqrt{2}}; \quad \frac{|a - a + s|}{\sqrt{2}} = \frac{|s|}{\sqrt{2}}.$$
This shows the symmetry we were interested in.

For the other planes that have parallel diagonals of opposite faces, the computation is similar.

In Section 5 we shall analyze the symmetries of each platonic solid — the cube in particular — from the viewpoint of rigid transformations. That method, which is much more powerful than the one used here, affords us an easier and more complete analysis.

Exercises

1. Give examples of figures with an infinite number of: (a) centers of symmetry, (b) axes of symmetry, (c) planes of symmetry.

2. Prove that if a figure in the Cartesian plane is symmetric with respect to both axes, then it necessarily has a center of symmetry.

3. Draw a curve $\mathcal{C}$ on the plane and fix a point $O \notin \mathcal{C}$. Complete the drawing so that the figure obtained is symmetric with respect to O. Repeat the exercise, now fixing a straight line $\mathcal{L}$.

4. Analyze the symmetries of all the conics, including the singular cases (that is, when the cutting plane passes through the vertex of the cone), and the symmetries of the quadric surfaces.

5. What characterizes the standard equation of a quadric surface which is symmetric with respect to a coordinate plane? What if we further assume it to be symmetric with respect to a coordinate axis? And with respect to the origin?

6. Prove that if a figure in the Cartesian space is symmetric with respect to each coordinate plane, then it is also symmetric with respect to the coordinate axes and to the origin.

7. How many axes of symmetry does an isosceles triangle have? And an equilateral triangle?

8. By making a drawing, determine the projections of $P(x, y, z)$ in the 3-space, onto each coordinate axis and each coordinate plane; draw the parallelepiped they determine.

1.2. Rigid transformations

9. Prove the Schwarz inequality by means of the inequality $||\bar{u} + t\bar{v}||^2 \geq 0$.

10. Justify formula (1.2) used to compute the distance from a point to a straight line in $\mathbb{R}^2$, and formula (1.3) used to obtain the distance from a point to a plane in $\mathbb{R}^3$.

11. Prove that the cross product of two vectors, $\bar{u}$, $\bar{v}$, is a vector orthogonal to both of them.

12. Prove that three vectors are coplanar if and only if their scalar triple product is zero.

13. Prove the following identities for vectors in $\mathbb{R}^3$, which imply that the cross product is not associative:
$$(\bar{u} \times \bar{v}) \times \bar{w} = (\bar{u} \cdot \bar{w})\bar{v} - (\bar{v} \cdot \bar{w})\bar{u},$$
$$(\bar{u} \times \bar{v}) \times \bar{w} + (\bar{v} \times \bar{w}) \times \bar{u} + (\bar{w} \times \bar{u}) \times \bar{v} = \bar{0}.$$

14. Give a formula (and justify it) to compute the distance between two **straight lines which cross one another** in a space (two straight lines cross one another in a space if they do not intersect and are not parallel).

15. Prove that the graph of the sine function is not symmetric with respect to the coordinate axes.

16. Complete the proof of the symmetries of a cone of revolution and make a drawing where all of them are shown.

17. Prove that the Euclidean axioms are satisfied in $\mathbb{R}^2$. (See these postulates in Appendix 5.4.)

18. Prove that a cube is symmetric with respect to each straight line joining midpoints of opposite faces.

19. Note that at each vertex of a cube, three edges meet. Use this property to convince yourself that the longest diagonals of a cube are not axes of symmetry.

20. Prove that the diagonals of a rectangle which is not a square, are not axes of symmetry of that rectangle. Are there any straight lines which are axes of symmetry of a rectangle? Is there any center of symmetry?

21. How many planes of symmetry does a cube have? And how many axes of symmetry? And how many centers? Justify your answers.

22. Answer the questions in Exercise 21 for the case of a sphere, and justify your statements.

1.2 Rigid transformations

At the beginning of this chapter, we stated that the mathematical concept which allows validating the process of superposition (used by Euclid to prove the congruence of two triangles) is the concept of a group of rigid transformations.

The concept of a group, which is very powerful and fundamental for many branches of mathematics, was first understood and used by Evariste Galois (1811–1832) around the year 1830 to decide on the solvability of equations of arbitrary degree by radicals. Later, in 1872, Felix Klein (1849–1925) gave a lecture known as the *Erlanger Programm* (because the university where Klein was going to work is located in Erlangen, a Bavarian city), in which he proposed to take the group concept as the axis of the study of geometry [Ke]:

Geometry is the study of invariants under a group of transformations.

For instance, the symmetry of a figure with respect to a plane can be expressed by saying that the figure remains invariant under a reflection in that plane. Other figures remain invariant when we rotate them about a point, such as a circle which rotates about its center, or when we translate them by a fixed distance, such as an infinite cylinder.

The reader will judge for himself or herself, throughout this book, the scope of Klein's proposal and the posterior development due to Sophus Lie. We recommend to the reader interested in knowing about the historical development of the concept of symmetry prior to the proposal by Klein and the works by Lie, the very pleasant book by Yaglom [Y].

The transformations allowed by Euclid are those which preserve the Euclidean distance, and we shall deal mainly with these. First, we look at the group corresponding to $\mathbb{R}^2$ and then the group of $\mathbb{R}^3$.

Definition. A **rigid transformation** of the plane is a surjective[1] function $T : \mathbb{R}^2 \to \mathbb{R}^2$ which preserves distances between points; that is,

$$d(P, Q) = d(T(P), T(Q)).$$

Rigid transformations are also called **isometries** because they preserve distances (and therefore measurements, such as areas, etc.)

Translations, rotations, and reflections are examples of rigid transformations on the plane. These transformations can be defined on the basis of intuitive situations.

Definition. A **translation on the plane by a fixed vector** $\bar{a} \in \mathbb{R}^2$ is the transformation $T_{\bar{a}} : \mathbb{R}^2 \to \mathbb{R}^2$ which moves each point by a distance equal to $||\bar{a}||$ in the direction of $\bar{a}$ (see Figure 1.13):

$$T_{\bar{a}}(P) = P + \bar{a}.$$

Notice that it was not necessary to use either the coordinates of P or those of $\bar{a}$; thus, the definition of translation in $\mathbb{R}^3$ will be the same. Also notice the difference in the notation of P and $\bar{a}$; this notation is due to the different roles they play: $\bar{a}$ is a fixed vector which moves each point $P \in \mathbb{R}^2$.

A translation is a rigid transformation, since

$$d(T_{\bar{a}}(P), T_{\bar{a}}(Q)) = ||P + \bar{a} - (Q + \bar{a})|| = ||P - Q|| = d(P, Q).$$

---

[1] It is not necessary to assume surjectivity; we do so for simplicity.

1.2. Rigid transformations

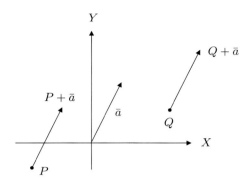

Figure 1.13: Translation in $\mathbb{R}^2$ by a fixed vector $\bar{a}$.

What can we say about the set of all translations?

To begin with, the composition of two translations, $T_{\bar{b}} \circ T_{\bar{a}}$, is another translation, namely $T_{\bar{a}+\bar{b}}$, as can be proved immediately by applying the composition to an arbitrary point P:

$$(T_{\bar{b}} \circ T_{\bar{a}})(P) = T_{\bar{b}}(T_{\bar{a}}(P)) = T_{\bar{b}}(P + \bar{a}) = P + \bar{a} + \bar{b} = T_{\bar{a}+\bar{b}}(P).$$

In addition, since the translation by the vector $\bar{0}$ leaves each point in its place (that is, $T_{\bar{0}}$ is the identity transformation), when composing $T_{\bar{0}}$ with any other translation $T_{\bar{a}}$ from the right or the left, we obtain $T_{\bar{a}}$ again:

$$T_{\bar{a}} \circ T_{\bar{0}} = T_{\bar{a}+\bar{0}} = T_{\bar{a}}.$$

Since a rigid transformation preserves distances, it is injective, or in other words it takes distinct points to distinct points.

This situation insures the existence of $(T_{\bar{a}})^{-1}$, but intuitively it is evident that to take each point back to its place after moving it by $\bar{a}$, it suffices to move it by $-\bar{a}$. Thus,

$$(T_{\bar{a}})^{-1} = T_{-\bar{a}}.$$

Moreover, since any composition of functions is associative, we have that

$$T_{\bar{c}} \circ (T_{\bar{b}} \circ T_{\bar{a}}) = (T_{\bar{c}} \circ T_{\bar{b}}) \circ T_{\bar{a}}.$$

All these properties mean that:

The set of translations of the plane forms a group under composition.

Before stating the definition of a group, which is fundamental throughout this book, it is important to emphasize the following fact:

Remark 2. Each vector of $\mathbb{R}^2$ determines a translation and, furthermore, *the sum of vectors determines the translation associated to the composition of the translations defined by those vectors.*

Now, $\mathbb{R}^2$ is a geometric object where we can measure distances and therefore we can speak of the concepts of *nearness* by means of *neighborhoods of radius ϵ*. On the other hand, we have proved that the transformations form a group. This is a very important property of $\mathbb{R}^2$ (and of any $\mathbb{R}^n$ in general): to have a geometric structure as well as an algebraic structure, and these structures "combine appropriately". This is what is known as a *Lie Group*. Throughout this book, we shall see that this same situation holds for other geometric objects which we know well.

Let us now give the definition of a group:

Definition. A **group** is a set G in which an operation is defined, that is, a function $\star : G \times G \to G$ with the following properties:

1a. *Closed:* The set G is closed under the operation $\star$. That is, the result of operating with two elements of G is an element of G:

$$g \star h \in G \text{ for all } g, h \in G.$$

Notice that this property is obvious from the definition of $\star$, since G is the codomain of this function, but this fact is so important that it is worthwhile emphasizing it.

2a. *Associativity:* For any three elements of G, it is the same to operate with the first two of them and then operate this result with the third one as to operate the first element with the result of operating with the last two of them. Namely,

$$(g \star h) \star k = g \star (h \star k).$$

3a. *Existence of the identity element:* There exists an element $e \in G$ such that when operating it with any other element of G the latter is not affected. With symbols,

$$e \star g = g \star e = g.$$

4a. *Existence of inverses:* For each $g \in G$, there exists another element of G, g^{-1}, such that when operating it with g the result obtained is the identity element e. With symbols,

$$g^{-1} \star g = g \star g^{-1} = e.$$

When additionally to these four properties *commutativity* also holds,

$$g_1 \star g_2 = g_2 \star g_1,$$

the group is called a **commutative or Abelian group**, after Niels Henrik Abel (1802–1829).

1.2. Rigid transformations

Several of the numerical systems the reader is familiar with, such as the integers, the rational numbers, and the real numbers, are groups. The integer numbers form a group, an Abelian group, with the operation of addition; the rational numbers and the real numbers are both Abelian groups under the operation of addition and, if zero is omitted (in both cases), the remaining numbers also form a commutative group under the operation of multiplication.

It is evident that translations of the plane form a commutative group under the operation of composition, and we know that vectors in $\mathbb{R}^2$ form a commutative group under the operation of addition. The correspondence is bijective and respects operations according to Remark 2; this is a very important property and it is worthwhile emphasizing it.

Definition. An **isomorphism of groups** is a bijective correspondence between two groups, $\phi : G \to G'$ which preserves group operations. That is:

$$\phi(g_1) \cdot \phi(g_2) = \phi(g_1 \star g_2)$$

for all g_1, g_2 in G, where $\star$ is the operation in G_1 and $\cdot$ is the operation in G'.

Thus one has:

The additive group $\mathbb{R}^2$ is isomorphic to the group of all translations of $\mathbb{R}^2$.

Another example of a group, which is fundamental in geometry, is that of the square $(n \times n)$-matrices with real entries and with determinant different from zero, $GL(n, \mathbb{R})$; the operation considered is the multiplication of matrices. We leave to the reader the proof of this statement for (2×2)-matrices as well as for (3×3)-matrices, which are the ones we shall use. Except for the case $n = 1$, these groups are **not** commutative, which is very easily proved.

In the Cartesian plane, the definitions of a rotation around the origin and a reflection with respect to a straight line through the origin, can be expressed in terms of (2×2)-matrices with real entries. This follows from the fact that the rotation of a point P by an angle θ around the origin in $\mathbb{R}^2$ can be reduced to the rotation of the vectors of a basis of $\mathbb{R}^2$, and similarly for the reflections. Let us see why this is so.

If we rotate the rectangular triangle corresponding to $P = x(1,0) + y(0,1)$, we obtain (see Figure 1.14)

$$Ro_\theta(P) = x \, Ro_\theta(1,0) + y \, Ro_\theta(0,1).$$

However, the rotation of $(1,0)$ by an angle θ about the origin transforms it to $(\cos\theta, \sin\theta)$, and takes $(0,1)$ to $(-\sin\theta, \cos\theta)$. Thus,

$$Ro_\theta(x,y) = x(\cos\theta, \sin\theta) + y(-\sin\theta, \cos\theta) = (x\cos\theta - y\sin\theta, x\sin\theta + y\cos\theta),$$

which justifies the following statement.

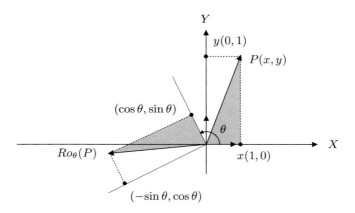

Figure 1.14: Rotation by an angle θ about the origin in $\mathbb{R}^2$.

Proposition. *The rotation by an angle θ around the origin in $\mathbb{R}^2$ is the function $Ro_\theta : \mathbb{R}^2 \to \mathbb{R}^2$ whose effect on a point $P(x, y)$ is*

$$Ro_\theta(x, y) = \begin{pmatrix} \cos\theta & -\sin\theta \\ \sin\theta & \cos\theta \end{pmatrix} \begin{pmatrix} x \\ y \end{pmatrix} = \begin{pmatrix} x\cos\theta - y\sin\theta \\ x\sin\theta + y\cos\theta \end{pmatrix}.$$

In other words, a rotation around the origin in the plane is the linear transformation determined by a matrix of the form $\begin{pmatrix} \cos\theta & -\sin\theta \\ \sin\theta & \cos\theta \end{pmatrix}$.

A rotation around the origin is a rigid transformation, since the rectangular triangles of Figure 1.14 are congruent and, accordingly, for any $P \in \mathbb{R}^2$, $||Ro_\theta(P)|| = ||P||$. Thus, taking into account the linearity of the transformation induced by the matrix above:

$$d(Ro_\theta(P), Ro_\theta(Q)) = ||Ro_\theta(P) - Ro_\theta(Q)|| = ||Ro_\theta(P - Q)|| = ||P - Q||.$$

Notice that the matrix of a rotation has the form

$$\begin{pmatrix} a & -b \\ b & a \end{pmatrix}, \text{ where } a^2 + b^2 = 1. \tag{1.4}$$

On the basis of this remark, it is very easy to prove that, as intuition suggests, when applying to a point $P(x, y)$ first the rotation by an angle θ and then the rotation by an angle ϕ, the effect is the same as rotating $P(x, y)$ by the angle $\theta + \phi$.

In fact, using the substitutions $a = \cos\theta$, $b = \sin\theta$, $c = \cos\phi$, $d = \sin\phi$ we get

$$\begin{pmatrix} c & -d \\ d & c \end{pmatrix} \begin{pmatrix} a & -b \\ b & a \end{pmatrix} = \begin{pmatrix} ca - db & -cb - ad \\ da + cb & -db + ca \end{pmatrix}.$$

Thus, the trigonometric identities for the cosine and the sine of the sum of two angles imply that $ca - db = \cos(\theta + \phi)$ and $da + cb = \sin(\theta + \phi)$. This proves that

1.2. Rigid transformations

the composition of two rotations $Ro_\phi \circ Ro_\theta$ is defined by the matrix corresponding to the rotation by an angle $\theta + \phi$.

That is, there is a bijective correspondence between the rotations about the origin by an angle $\theta \in [0, 2\pi)$ and the matrices of the form (1.4). Moreover, that correspondence respects the operations, namely, the composition of rotations corresponds to the multiplication of matrices.

By taking into account this correspondence, it is easy to prove the following theorem.

Theorem. *Rotations around the origin form a commutative group.*

Proof. We prove each of the conditions for having a group structure.

Closed: The composition of two rotations is another rotation, as we have already seen.

Associativity: The composition of rotations is associative, since that is true for any functions. Nevertheless, in this case it also follows from the fact that the multiplication of matrices is associative, as the reader shall prove in Exercise 1.

Existence of the identity element: The rotation by an angle 0 corresponds to the matrix

$$\begin{pmatrix} 1 & 0 \\ 0 & 1 \end{pmatrix},$$

which is the identity element for the multiplication of matrices.

Each rotation has an inverse rotation: When composing the rotation by the angle $-\theta$ with the rotation by the angle θ, the result is the rotation by the angle 0, which is the identity element for the composition of rotations, for if $a = \cos\theta$ and $b = \sin\theta$,

$$\begin{pmatrix} a & b \\ -b & a \end{pmatrix} \begin{pmatrix} a & -b \\ b & a \end{pmatrix} = \begin{pmatrix} a^2 + b^2 & -ab + ab \\ -ba + ab & b^2 + a^2 \end{pmatrix} = \begin{pmatrix} 1 & 0 \\ 0 & 1 \end{pmatrix}.$$

Commutativity: The composition of two rotations can be applied in either order without affecting the result. The proof of this property is left to the reader. □

Notice that these arguments actually show that the set of all real matrices of the form (1.4) is an Abelian group, isomorphic to the group of rotations around the origin in $\mathbb{R}^2$.

Our last example of a rigid transformation on the plane is that of reflections with respect to a straight line through the origin. For this example we also give a definition in agreement with intuition; this definition implies that these reflections are linear transformations of the plane.

In Figure 1.15, we have again associated to the point P the rectangular triangle corresponding to $P = x(1, 0) + y(0, 1)$, and we have reflected the whole triangle with respect to the straight line $\mathcal{L}_\phi$ (think of it as if it were a mirror), which forms an angle ϕ with the positive X axis, denoted as X^+.

The reflected triangle is again a rectangular triangle, and the corresponding vector $Re_\phi(P)$ can be formed as follows:

$$Re_\phi(P) = x\,Re_\phi(1,0) + y\,Re_\phi(0,1).$$

Since the straight line forms the angle ϕ with X^+, the reflection of $(1,0)$ with respect to the straight line $\mathcal{L}_\phi$ forms the angle 2ϕ with X^+ and, accordingly, its coordinates are $(\cos 2\phi, \sin 2\phi)$.

The reflection of $(0,1)$ does not form an angle of $\pi/2$ with $(cos2\phi, sin2\phi)$, but an angle of $-\pi/2$ because a reflection changes the orientation of the angles. Hence, the reflection of $(0,1)$ is $(\sin 2\phi, -\cos 2\phi)$, and we are ready to construct the matrix that gives the reflection.

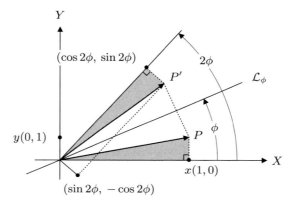

Figure 1.15: Reflection with respect to a straight line through the origin.

Proposition. *The* **reflection with respect to the straight line** $\mathcal{L}_\phi$ *through the origin, which forms an angle ϕ with the positive X axis on the plane, is the transformation $Re_\phi : \mathbb{R}^2 \to \mathbb{R}^2$ whose effect on a point $P(x,y)$ is*

$$Re_\phi(x,y) = \begin{pmatrix} \cos 2\phi & \sin 2\phi \\ \sin 2\phi & -\cos 2\phi \end{pmatrix} \begin{pmatrix} x \\ y \end{pmatrix} = \begin{pmatrix} x \cos 2\phi + y \sin 2\phi \\ x \sin 2\phi - y \cos 2\phi \end{pmatrix}.$$

The proof that a reflection of this type is a rigid transformation is left to the reader.

This time the matrix has the form

$$\begin{pmatrix} a & b \\ b & -a \end{pmatrix}, \text{ where } a^2 + b^2 = 1. \tag{1.5}$$

Notice that the determinant of a matrix of a rotation is 1, while the determinant of a matrix of a reflection is -1. These results imply that the composition of two reflections can not be a reflection, for the corresponding matrices are multiplied

1.2. Rigid transformations

and the determinant of the matrix of their product is equal to the multiplication of their respective determinants: 1.

Therefore, we can not think of proving that the reflections form a group, since the product of two of them is not a reflection anymore. Also, the identity element of the multiplication of matrices has determinant 1, so it is not a reflection.

However, it is true that the inverse of a reflection is itself a reflection, as intuition suggests: If P' is the reflection of a point P with respect to a straight line, and then P' is reflected with respect to that same straight line, we obtain P again (see Exercise 5 below). We invite the reader to prove this result formally by multiplying by itself the matrix of a reflection.

Now, if we write the matrix of a rotation in the form (1.4), and that of a reflection in the form (1.5), we shall be able to prove easily that their product is a reflection, although the straight line with respect to which the reflection is done depends on the order of multiplication (see Exercise 5). In other words, the composition of a rotation with a reflection (in either order) is again a reflection.

This suggests the following theorem.

Theorem. *The set of rotations in $\mathbb{R}^2$ about $(0,0)$ and reflections with respect to a straight line through $(0,0)$ form a group under composition.*

The proof is left to the reader. Just remember that the inverse of a multiplication of matrices (or of the composition of transformations) is the multiplication of the inverses *taken in reverse order*.

This group is called the **orthogonal group of order 2**, $O(2, \mathbb{R})$, because the associated matrices have the elements of an orthonormal basis as column vectors: a right basis for the rotations and left one for the reflections.

For the matrices M whose column vectors constitute an orthonormal basis of $\mathbb{R}^2$ (or of $\mathbb{R}^n$), it is immediately verified that their inverse, M^{-1}, is equal to their **transpose**, M^t, the matrix whose columns are the rows of the original matrix (respecting their order), and they are called **orthogonal matrices.**

The orthogonal matrices whose determinant is $+1$, which simply is the group of rotations, form the so-called **special orthogonal group**, which is denoted as $SO(2, \mathbb{R})$.

Before beginning with the study of rigid transformations in $\mathbb{R}^3$, let us remember that, since in $\mathbb{R}^3$ we have a way of measuring the distance between $\bar{a}$ and $\bar{b}$, the isomorphism between the additive group $\mathbb{R}^2$ and the group of translations of the plane (under composition) allows us to speak of "nearby" translations, or more precisely of the neighborhood of radius ϵ of a translation $T_{\bar{a}}$: it is formed by all the translations $T_{\bar{b}}$ such that $||\bar{a} - \bar{b}|| < \epsilon$.

An analogous property holds for the group of rotations around the origin: we can identify the rotation by an angle θ with the point of the unit circle in $\mathbb{R}^2$:

$$S^1 = \{(x,y) \in \mathbb{R}^2 \,|\, x^2 + y^2 = 1\}$$

determined by the oriented line through the origin that forms an angle θ (measured counter-clock-wise) with the the positive X axis. We can say how close the points of the circle are because we know how to measure distances in S^1. Hence we can also measure how close two rotations of the plane are. In fact S^1 can be seen as the multiplicative group of the complex numbers of norm 1, and the correspondence with $SO(2, \mathbb{R})$ is a group isomorphism.

Sophus Lie (1842–1899) was the first mathematician to study this type of groups. He was a contemporary and friend of Klein. These groups are called **Lie groups** after him. We shall study them further on in this book and, in the case of rotations, we shall show the convenience of using complex coordinates.

As to rigid transformations of the three-dimensional Euclidean space into itself, the space we live in, we can also give as examples translations, rotations, and reflections.

Translations are defined exactly the same way as in the case of the plane, for the **translation in $\mathbb{R}^3$ by the fixed vector** $\bar{a} \in \mathbb{R}^3$ is given by

$$T_{\bar{a}} : \mathbb{R}^3 \to \mathbb{R}^3 \text{ such that } T_{\bar{a}}(P) = P + \bar{a}.$$

Notice that, in the proof that the translations in $\mathbb{R}^2$ constitute a group, the fact that $\bar{a}$ and P are elements of $\mathbb{R}^2$ was not really used; hence the same arguments imply that translations in $\mathbb{R}^3$ form a commutative group. Similar statements hold also for the corresponding groups of rigid transformations, and the proof of rigidity also applies in this case.

In fact the previous discussion about translations can be fully generalized to $\mathbb{R}^n$, and therefore every element $\bar{a} \in \mathbb{R}^n$ can be regarded as a translation, a rigid transformation of $\mathbb{R}^n$.

To define a rotation and a reflection in $\mathbb{R}^3$, we propose extending the corresponding definitions in $\mathbb{R}^2$.

Let us remember that the standard basis of $\mathbb{R}^3$ is formed by the vectors $\hat{e}_1 = (1, 0, 0)$, $\hat{e}_2 = (0, 1, 0)$, and $\hat{e}_3 = (0, 0, 1)$, the three of them with norm 1, being orthogonal one to another, and forming a right basis.

As for a rotation in $\mathbb{R}^2$, we shall require a right basis to transform into another right basis. The reader may think of three rods welded together at one of their extremes, so that they remain perpendicular to one another; they can move but their common extreme must remain fixed. Hence, it is possible to define a rotation in $\mathbb{R}^3$ by means of a 3×3 orthogonal matrix with determinant 1.

Definition. A **rotation about the origin in $\mathbb{R}^3$** is a linear transformation $Ro : \mathbb{R}^3 \to \mathbb{R}^3$ determined by a matrix of the form

$$\begin{pmatrix} u_1 & v_1 & w_1 \\ u_2 & v_2 & w_2 \\ u_3 & v_3 & w_3 \end{pmatrix},$$

where $\hat{u} = (u_1, u_2, u_3)$, $\hat{v} = (v_1, v_2, v_3)$, $\hat{w} = (w_1, w_2, w_3)$ form a right orthonormal basis.

1.2. Rigid transformations

It is clear that under a rotation around $\bar{0}$ in $\mathbb{R}^3$, any sphere with that center is transformed into itself, particularly the sphere of radius 1,

$$S^2 = \{(x, y, z) \in \mathbb{R}^3 \mid x^2 + y^2 + z^2 = 1\}.$$

Our experience of playing with a ball shows us that if we want to make the ball rotate into our hands, we must place one hand opposite to the other. That corresponds to the algebraic fact that a rotation about the origin in $\mathbb{R}^3$ leaves a certain straight line through the origin fixed point to point, which is the **axis of rotation**.

That is so because the characteristic polynomial is of the third degree and it must have a real root. With that characteristic value (proper value or eigenvalue), we determine an invariant subspace of the rotation. Accordingly, any rotation in $\mathbb{R}^3$ around the origin reduces to a rotation on the plane that is perpendicular to the axis of rotation (see [Ra]).

For a reflection, we require the standard basis to transform into a **left orthonormal basis**.

Definition. A **reflection about the origin in $\mathbb{R}^3$** is a linear transformation $Re : \mathbb{R}^3 \to \mathbb{R}^3$ determined by a matrix of the form

$$\begin{pmatrix} u_1 & v_1 & w_1 \\ u_2 & v_2 & w_2 \\ u_3 & v_3 & w_3 \end{pmatrix},$$

where $\hat{u} = (u_1, u_2, u_3)$, $\hat{v} = (v_1, v_2, v_3)$, $\hat{w} = (w_1, w_2, w_3)$ form a left orthonormal basis.

The following are some examples of reflection matrices in $\mathbb{R}^3$:

$$\begin{pmatrix} -1 & 0 & 0 \\ 0 & a & -b \\ 0 & b & a \end{pmatrix}, \quad \begin{pmatrix} 1 & 0 & 0 \\ 0 & a & b \\ 0 & b & -a \end{pmatrix} \quad \text{and} \quad \begin{pmatrix} -1 & 0 & 0 \\ 0 & -1 & 0 \\ 0 & 0 & -1 \end{pmatrix}.$$

The first matrix leaves invariant the X axis, although it exchanges the semi-axes (on the YZ plane we have a rotation). The second matrix fixes point by point the X axis and makes a reflection on the YZ plane. The third matrix corresponds to a very interesting case: the **antipodal** application, which leaves invariant all the straight lines through the origin, although each point is transformed into its symmetric image with respect to the origin.

Every sphere with the center at the origin is mapped into itself under a reflection. If we investigate the invariant subspaces under reflection, we find again a straight line through the origin, for the characteristic polynomial is of degree 3, although in this case the points may not remain fixed, but exchange with those of the complementary half straight line.

The proof that rotations and reflections in $\mathbb{R}^3$ are rigid transformations is based again on the fact that they are linear transformations and, since we do not use coordinates as in the case of the plane, that same proof holds for $\mathbb{R}^3$.

Again, it happens that rotations and reflections in $\mathbb{R}^3$ constitute a group under composition (the composition corresponds to the operation of matrix multiplication). In this case, the group is denoted as $O(3,\mathbb{R})$, and it is called the **orthogonal group of order 3**.

The definition of rotations and reflections in $\mathbb{R}^3$ as linear transformations makes the origin remain fixed. However, it is clear that we can apply rotations and reflections with respect to any point of $\mathbb{R}^3$. The treatment we have provided may seem restrictive, but the following considerations will show that is not so.

Proposition. *Any rigid transformation is the composition of a translation with an orthogonal transformation.*

Proof. Let us denote with T a rigid transformation in $\mathbb{R}^3$. If $T(\bar{0}) = \bar{a}$, the composition of $T_{-\bar{a}}$ with T, leaves $\bar{0}$ fixed. Now it will suffice to prove that any rigid transformation U which leaves $\bar{0}$ fixed is orthogonal. The details are left to the reader.

A possible way of proving this claim is to show that:

(i) U respects norms, that is, $||\bar{a}|| = ||U(\bar{a})||$.

(ii) U respects the scalar product. For proving it, we can develop both members of the equality
$$||\bar{a} - \bar{b}||^2 = ||U(\bar{a}) - U(\bar{b})||^2,$$
which holds because U is rigid, and take into account (i).

(iii) U is linear. For proving it, verify that $||U(\lambda \bar{a}) - \lambda U(\bar{a})||^2 = 0$, and $||U(\bar{a} + \bar{b}) - U(\bar{a}) - U(\bar{b})||^2 = 0$.

Since $\bar{0}$ is the only vector characterized by its norm, that proves that U "takes out" scalars and distributes over addition. Hence, the matrix corresponding to U in an orthonormal basis is orthogonal. $\square$

It is easy to convince oneself that any rigid transformation on the plane is determined by three non-collinear points and their images, and that any rigid transformation in three-dimensional space is determined by four non-coplanar points and their images (and so on).

Considering these results, it can be proved that any rigid transformation on the plane is the product of at most three reflections (see Exercise 7 and Exercise 8 below).

The group of rigid transformations of the plane on the plane is the **Euclidean group of order 2**, $E(2)$, and the group corresponding to $\mathbb{R}^3$ is the **Euclidean group of order 3**, $E(3)$.

1.2. Rigid transformations

In Section 1.3 we show that we are familiar with many examples of invariance under rigid transformations. Moreover, we shall introduce a fundamental concept of differential geometry, that of curvature.

Exercises

1. Is $\mathbb{Z} - \{0\}$ a group with respect to multiplication?

2. Prove that $GL(2, \mathbb{R})$, the set of 2×2 matrices with real entries and determinant different from zero, forms a group under multiplication called the **general linear group of order 2**. Is this group commutative?

3. Prove that the multiplication of matrices of the type (1.4) is commutative.

4. Prove that a reflection on the plane with respect to a straight line through the origin is a rigid transformation.

5. Prove that the inverse transformation of a reflection of the plane with respect to one line through the origin is that same reflection. That is, given a reflection $Re : \mathbb{R}^2 \to \mathbb{R}^2$, one has that $Re \circ Re$ is the identity.

6. Determine the line of reflection that corresponds to the composition of a rotation and a reflection; verify that this line depends on the order of the composition.

7. Prove that any rotation on the plane about the origin is the composition of two reflections with respect to straight lines through the origin.

8. Prove that any translation on the plane is the composition of two reflections with respect to parallel straight lines which are perpendicular to the direction of the translation.

9. Prove that the statements in Exercises 7 and 8 can be generalized this way:
   "The composition of two reflections with respect to arbitrary straight lines is a rotation, except in the case when the straight lines are parallel, where a translation is obtained."
   Therefore, it makes sense to state that a translation is the limit of rotations. Can you justify this latter statement?

10. Prove that a rigid transformation on the plane is determined when the images A', B', and C' of three non-collinear points A, B, and C are known.

11. Prove that any rigid transformation on the plane is the composition of at most three reflections. For that, take three points A, B, and C and their images A', B', and C', and verify that the statement holds for each of the following cases, which are all possible cases.
    i) $A = A'$, $B = B'$ and $C = C'$;
    ii) $A = A'$, $B = B'$ and $C \neq C'$;
    iii) $A = A'$, $B \neq B'$ and $C \neq C'$;
    iv) $A \neq A'$, $B \neq B'$ and $C \neq C'$.

12. Establish for $\mathbb{R}^3$ the statement analogous to that in Exercise 6 and prove it.

13. Establish for $\mathbb{R}^3$ the statement analogous to that in Exercise 7 and prove it.

14. Establish for $\mathbb{R}^3$ the statement analogous to that in Exercise 10 and prove it.

15. Associate to each 3×3 matrix M with real entries an element of $\mathbb{R}^9$ writing the rows one next to the other, and reciprocally. Define in $\mathbb{R}^9$ a distance by means of the scalar product, and use it to define a distance between the 3×3 matrices. Prove that if the distance d between matrices is restricted to the elements of $O(3,\mathbb{R})$, for any $M \in SO(3,\mathbb{R})$, there exist $\epsilon > 0$ such that if $d(M, M') < \epsilon$, then M' can not be a reflection. (This says, in other words, that $SO(3,\mathbb{R})$ is an open subset of $O(3,\mathbb{R})$. In fact $O(3,\mathbb{R})$ has two disjoint "pieces", or connected components, and $SO(3,\mathbb{R})$ is one of these.)

16. Prove that there is an isomorphism between the field $\mathbb{C}$ of complex numbers $x + iy$ and the set of matrices with real entries of the form

$$\begin{pmatrix} x & -y \\ y & x \end{pmatrix},$$

provided by the usual addition and multiplication for matrices. Prove as well that those matrices are the composition of a rotation with a **homothesis** $H_\rho : \mathbb{R}^2 \mapsto \mathbb{R}^2$ that takes (x,y) into $(\rho x, \rho y)$ with $\rho = \|(x,y)\|$.

1.3 Invariants under rigid transformations

Now we analyze what properties of geometrical objects are preserved — that is, what properties are invariant — under rigid transformations.

Of course, the types of triangles — equilateral, isosceles, and scalene — are invariant characterizations under rigid transformations, since they are defined in terms of distances. If the three sides of a triangle ABC are equal (in length), the same holds for the triangle $A'B'C'$ obtained by applying a rigid transformation to the triangle ABC. Moreover, the lengths of the sides are preserved and that forces the angles to have the same measurements. (Can you give a justification of this assertion?) The analogous properties hold for the other two types of triangles.

Parallel straight lines are taken into parallel straight lines under a rigid transformation, because parallel straight lines are equidistant; and straight lines that intersect one another at a certain angle are transformed into other straight lines that intersect one another with that same angle (consider the triangle formed by one point on each straight line and its point of intersection). Therefore, under a rigid transformation a square is transformed into another square, and so on.

The types of conic sections are also invariant under rigid transformations. For instance, an ellipse is the locus of the points P on the plane such that the sum of their distances to two fixed points F_1 and F_2, called foci, remains constant (which is usually denoted as $2a$). Thus, if the images of the foci under a rigid transformation T are F_1' and F_2', then the defining condition of an ellipse holds

1.3. Invariants under rigid transformations

for the point $P' = T(P)$ with respect to F_1' and F_2'. Moreover, of course, the semi-axes of the latter ellipse will have the same length as those of the semi-axes of the former.

The same reasoning is applied to the other types of conic sections and to quadric surfaces in general. To explain this, let us start by making precise what is a quadric surface and how they are classified.

A **quadric surface** is the locus of the points (x, y, z) which satisfy a second degree equation in three variables,

$$Ax^2 + By^2 + Cz^2 + 2Dxy + 2Exz + 2Fyz + Gx + Hy + Iz + J = 0,$$

and the different types of quadric surfaces, a total of fifteen (some of them degenerated), are obtained by varying the coefficients of this equation.

The classification of the different types of quadric surfaces is made, first, on the basis of the **matrix of the quadratic form** determined by the symmetric matrix

$$\begin{pmatrix} A & D & E \\ D & B & F \\ E & F & C \end{pmatrix}.$$

The reader may recall that a matrix like this one is always diagonalizable and the entries in its diagonal, the characteristic values of the transformation corresponding to the matrix, indicate how much the characteristic vectors are dilated or contracted, and if their direction is unchanged or reversed (see [B-ML] or [Ra]).

The **rank of a matrix** is the dimension of its image, which in this case is the same as the number of its characteristic values different from zero; its **signature** is the difference between the number of its positive proper values and that of its negative proper values. The rank and the signature of a matrix are invariant under orthogonal transformations because the characteristic polynomial is invariant.

The invariance of a quadric surface (concerning its type and measure) under a rigid transformation will be proved by the following facts:

1º. The degree of a polynomial is invariant under rigid transformations, for it is invariant under a translation as well as under an orthogonal transformation. Accordingly, a quadratic polynomial is transformed into another quadratic polynomial, that is, every quadric surface is transformed into another quadric surface.

2º. The only standard forms of quadric surfaces are 1), 3), 7) and 13) in the list below. They are shown in Figure 1.16).

- Rank 3:
  1) The empty set, which corresponds to the equation $x^2 + y^2 + z^2 = -1$.
  2) An ellipsoid, which corresponds to the equation $x^2 + 2y^2 + 3z^2 = 1$.
  3) A point, given by the equation $x^2 + 2y^2 + 3z^2 = 0$.

4) A two-sheeted hyperboloid, given by the equation $x^2 - 2y^2 - 3z^2 = 1$.
5) A one-sheeted hyperboloid, with typical equation is $x^2 + 2y^2 - 3z^2 = 1$.
6) A cone, whose typical equation is $x^2 + 2y^2 - 3z^2 = 0$.

- Rank 2:
  *) The empty set, given by $x^2 + y^2 = -1$.
  7) A straight line, given by the equation $x^2 + 2y^2 = 0$.
  8) Two intersecting planes, whose equation is $x^2 - 2y^2 = 0$.
  9) An elliptic cylinder, whose typical equation is $x^2 + 2y^2 = 1$.
  10) A hyperbolic cylinder, given by the equation $x^2 - 2y^2 = 1$.
  11) A hyperbolic paraboloid, whose equation is $x^2 - 2y^2 = z$.
  12) An elliptic paraboloid, with equation $x^2 + 2y^2 = z$.

- Rank 1:
  **) The empty set, given by $x^2 = -1$.
  13) A double plane, whose typical equation is $x^2 = 0$.
  14) Two parallel planes, which correspond to the equation $x^2 = 1$.
  15) A parabolic cylinder, whose typical equation is $x^2 = y$.

3º. There exist quadric surfaces with identical rank and signature but distinct type. The difference is due to the non-quadratic part in their equation, which makes them have distinct properties, as for instance regarding their dimension, or their **extension**: whether or not they are **bounded** (when the surface is contained in some ball with center at the origin), or whether the surface is limited by some plane (as occurs with a paraboloid), or it is **ruled**, (if through each of their points passes a straight line completely contained in the surface), if the surface is constituted by only one piece, and so on.

This last part is left to the reader as an exercise, following the reasoning we give in the case of quadric surfaces of rank 3.

- Signature 3: The two surfaces of rank 3 with signature 3 are an ellipsoid and a point, but the independent term, which is not affected by a rigid transformation, gives two degrees of freedom to the points that satisfy (2) and no degree of freedom to the only point that satisfies (3).

- Signature 1: The two surfaces of rank 3 with signature 1 are a hyperboloid of one sheet and a cone, but the independent term distinguishes them: if in equation (5) we leave in the left-hand member a difference of squares, the new right-hand member is a difference of squares as well, and that shows that we have a double-ruled surface (two straight lines completely contained in the hyperboloid pass through each point), while a cone is a simply ruled surface (only one straight line completely contained in the cone passes through each point different from the vertex: the generatrix to which that point belongs).

- Signature -1: There exists only one surface of rank 3 and signature -1: the hyperboloid of two sheets.

1.3. Invariants under rigid transformations

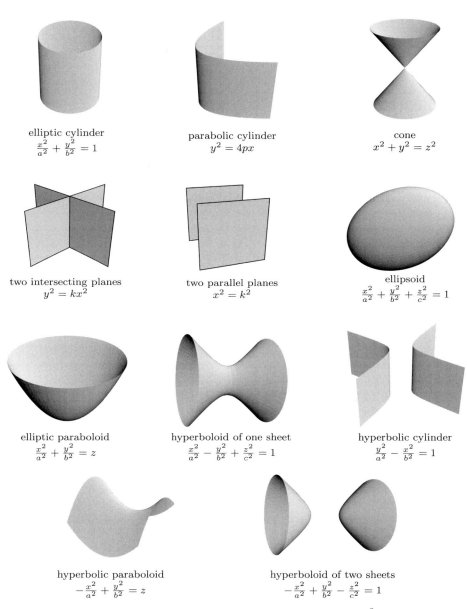

Figure 1.16: Non-degenerated quadric surfaces in $\mathbb{R}^3$.

It is clear that not all of the loci listed above can be called surfaces, such as the empty set (equations 1,) *), and **)), the point (3), and the straight line (7). These are limit cases of an ellipsoid and an elliptic cylinder, respectively; thus, they are called **degenerated quadrics**. However, all of them are loci that correspond to quadratic polynomial equations in three variables.

The proof of the invariance of the rank and the signature of a matrix under a non-singular transformation can be studied in [B-ML]. A consequence of that invariance is the invariance of the degree.

Now we give a first version of the most important invariant under isometries in differential geometry: the **curvature of a surface at each of its points**; as we shall see, the value of the curvature may change from point to point. A good reference for those who wish to learn more about this topic is [DoC].

The figures studied in differential geometry must be **smooth**. That is, if the figures under study are curves, it is required that the tangent line at each of their points be well-defined; if the figures under study are surfaces, the condition is that the tangent plane at each of their points be well-defined.

We recall that a smooth curve has a well-defined tangent line at each of its points. This allows us to approximate the curve by these tangent lines in a **sufficiently small neighborhood** of each point, which is very useful. Analogously, a surface has its tangent plane at each of its points, giving us the possibility of substituting the surface by its tangent plane locally (see Figure 1.17).

The latter situation does not always occur: there can be points where the surface fails to have a well-defined tangent space. For instance consider a surface as simple as a cone of revolution; the tangent plane can not be defined at its vertex. To convince ourselves of that, it suffices to consider that each of the generating lines is not just tangent to the cone, but it is also contained in it. If there existed a tangent plane to the cone at its vertex, each generatrix should be contained in that plane, but this is evidently impossible.

Calculus provides a simple technique to detect when a surface is smooth at a given point. This is simpler when the surface under study is defined by a differentiable function $F : \mathbb{R}^3 \to \mathbb{R}$, such as $E(x,y,z) = x^2 + 4y^2 + 9z^2$, $P(x,y,z) = x^2 + y^2 - z$, $\Pi(x,y,z) = x + 2y + 3z$, etc. (To clarify some concepts mentioned in what follows in this section, the reader may consult our Appendix 5.1, or also Appendix 5.1 of [Cou].)

A hyperboloid of two sheets (4), a cone (6), and a hyperboloid of one sheet (5), are all **level surfaces** of the function

$$F(x,y,z) = x^2 - y^2 - z^2,$$

whose gradient ∇F is

$$\nabla F(x,y,z) = (2x, -2y, -2z),$$

which becomes zero only at $(0,0,0)$.

1.3. Invariants under rigid transformations

The origin does not belong to the hyperboloid of two sheets $x^2 - y^2 - z^2 = 1$ (nor to the hyperboloid of one sheet) and thus $\nabla F(P) \neq (0,0,0)$ at every point P of the hyperboloid. That is why we say that 1 is a **regular value** of F.

On the other hand, the origin does belong to the cone $x^2 - y^2 - z^2 = 0$, being its vertex, and since $\nabla F(0,0,0) = (0,0,0)$, we say that 0 is a **critical value** of F.

If a surface S is the **inverse image of a regular value** r (see Appendix 5.1) of a differentiable function $F : \mathbb{R}^3 \to \mathbb{R}$, that is, if

$$S = \{(x,y,z) \in \mathbb{R}^3 \mid F(x,y,z) = r \text{ and } \nabla F(x,y,z) \neq (0,0,0)\},$$

then the implicit function theorem implies that S is smooth at each point, having everywhere a well-defined tangent space. In fact it is easy to prove, using the Chain Rule, that for any $P_0 \in S$, the gradient of F at the point P_0, $\nabla F(P_0)$, is a vector perpendicular to the velocity vector $\alpha'(t_0)$ of every smooth curve $\alpha(t) = (x(t), y(t), z(t))$ contained in S and passing through P_0 at the instant t_0. Let us see how the curve is contained in S,

$$F(x(t), y(t), z(t)) = r \text{ for every } t \in \text{Dom}(\alpha).$$

By deriving with respect to t, evaluating at t_0 and by using the Chain Rule we obtain

$$\frac{dF}{dt}(t_0) = \nabla F(P_0) \cdot \alpha'(t_0) = 0.$$

Hence, the plane through P_0 and perpendicular to $\nabla F(P_0)$ contains the tangent vector of every smooth curve contained in S through P_0; that is why it is called the **tangent plane** to S at P_0. Its equation is

$$(P - P_0) \cdot \nabla F(P_0) = 0. \tag{1.6}$$

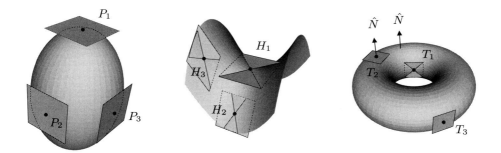

Figure 1.17: Tangent planes to different surfaces at some of their points.

In Figure 1.17 we illustrate several planes tangent to an ellipsoid, a hyperboloid, a paraboloid and a **torus of revolution**.

In the case of the ellipsoid, the plane tangent at one of its points intersects it only at that point, and leaves the whole surface only on one side of the tangent plane. Points of this type are called **elliptic points**.

In the case of the hyperbolic paraboloid, the plane tangent at one of its points intersects it in two straight lines, since it is a double-ruled surface, and there are points of the surface at both sides of the tangent plane. The latter is a necessary condition for a point to be a **hyperbolic point**.

In the case of the torus of revolution, the intersection with the tangent plane depends on where the point is located. There are elliptic points, hyperbolic points, and **parabolic points** in the torus, say $\mathbb{T}$, for which there exists a curve $\mathcal{C}$ such that at every point in $\mathcal{C}$, the vector normal to $\mathbb{T}$ is fixed and the tangent plane leaves $\mathbb{T}$ in one of the two semi-spaces it defines.

The reader may verify these statements by determining the equation of the tangent plane with formula (1.6), for all the aforementioned surfaces are inverse images of regular values of differentiable functions from $\mathbb{R}^3$ in $\mathbb{R}$ (see Exercise 6c).

Notice that given a surface S defined as above, i.e., the inverse image of a regular point of smooth function F in $\mathbb{R}^3$, at each point $P \in S$ the vector $\nabla F(P)$ is normal to S; this vector spans a real line, and each plane in $\mathbb{R}^3$ containing this line intersects the surface in a curve called a **normal section**. In Figure 1.18 we show normal sections $\mathcal{C}_1$, $\mathcal{C}_2$ and $\mathcal{C}_3$ at the point P_0 of a cylinder, an ellipsoid, and a hyperbolic paraboloid.

If a smooth curve $\alpha(s) = (x(s), y(s), z(s))$ is parametrized so that its tangent vector $\alpha'(s)$ always has norm 1, we can define its **curvature** $k(s_0)$ at the point $\alpha(s_0)$ as the norm of the vector $\alpha''(s_0) = (x''(s_0), y''(s_0), z''(s_0))$.

This is so because $||\alpha'(s)||$ is constant, and thus by obtaining the derivative of this vector we only measure the *variation with respect to its position*, the speed at which the curve gets away from the tangent line at the point.

The parameter s with which we obtain $||\alpha'(s)||$ is called **arc length**, for it allows us to traverse equal arcs at equal time intervals. Any smooth curve admits this type of parametrization (see [DoC]).

For a straight line, the curvature at any of its points is zero, since the parametrization of a straight line by arc length is $\alpha(s) = P_0 + s\widehat{u}$, given P_0 and $\widehat{u}$ constants, with $||\widehat{u}|| = 1$, it turns out that $\alpha''(s) = \bar{0}$ for all s.

For a circle, the curvature is the inverse of the radius, since if the parametrization is

$$\alpha(s) = (r\cos(s/r), r\sin(s/r)),$$

then the velocity vector is

$$\alpha'(s) = (-\sin(s/r), \cos(s/r)),$$

whose norm is 1, and the acceleration vector is

$$\alpha''(s) = (-(1/r)\cos(s/r), -(1/r)\sin(s/r)),$$

1.3. Invariants under rigid transformations

whose norm is $1/r$. Moreover, in all cases it points toward the concave part of the circle.

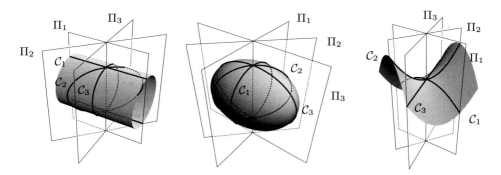

Figure 1.18: Normal sections of a cylinder, an ellipsoid, and a hyperbolic paraboloid.

The definition of **curvature of a surface** $S \subset \mathbb{R}^3$ **at one of its points** P_0 requires giving a sign to the curvatures of the normal sections. For that purpose, we fix one of the possible unit normal vectors, $\widehat{N}(P_0)$, at the point and project the vector $\alpha''(s_0)$ on it; if the projection has the same direction as $\widehat{N}(P_0)$, the section will have positive curvature, and negative if the projection has the reversed direction of $\widehat{N}(P_0)$.

In the case of an ellipsoid, the acceleration vectors of all normal sections will be on the same side of the tangent plane, while in the case of a hyperbolic paraboloid, at each point there will be normal sections on both sides of the tangent plane, as occurs with the curves $\mathcal{C}_1$ and $\mathcal{C}_3$ of the hyperbolic paraboloid of Figure 1.18.

There are as many normal sections as diameters in a small circle centered at P_0 of the tangent plane, in other words, one for each $\theta \in [0, \pi]$. Therefore, the curvature of a normal section, which is called **sectional curvature**, can be seen as a function of the angle θ, $k(\theta)$.

A very important result in calculus establishes that a continuous function defined in a closed interval takes its maximum and its minimum. Therefore, for some $\theta_1, \theta_2 \in [0, \pi]$, we have $k_1 = k(\theta_1)$ as the **maximal curvature** and $k_2 = k(\theta_2)$ as the **minimal curvature** of the normal sections.

The **curvature of a surface at one of its points**, $K(P)$, was defined by Leonhard Euler (1707–1783) as the product of the maximal and minimal curvatures

$$K(P) = k_1(P) k_2(P).$$

Notice that the value of $K(P)$ does not change even if we choose as normal vector $-\widehat{N}(P_0)$. At present, $K(P)$ is called **Gaussian curvature** of the surface at a point P, after Karl Friedrich Gauss (1777–1865), who identified this as one of the most important concepts in differential geometry (see [DoC]).

When $K(P)$ is constant, the surface is called a **surface of constant curvature**. A sphere has constant curvature $K = 1/r^2$, where r is the length of its radius, and a plane has constant curvature $K = 0$.

Since the norm of $\alpha''(s)$ is invariant under rigid transformations, when applying a rigid transformation to a surface, the curvatures of the normal sections are also invariant.

In Exercise 5 we leave the reader the task of completing the proof of the invariance of the curvature $K(P)$ under rigid transformations, and of computing the curvature of the surfaces at the points indicated in the Exercises. The following formula will be useful; it expresses the curvature $k(t)$ of a curve $\alpha(t)$ with an arbitrary parameter (see Exercise 5):

$$k(t) = \frac{||\alpha' \times \alpha''||}{||\alpha'||^3}. \tag{1.7}$$

We suggest to examine the image of the surface to decide which are the sections of maximal and minimal curvature.

Before finishing this section, we define two other concepts which can be easily understood if we analyze them in the examples of the surfaces we have been working with.

The first concept is that of a **homogeneous surface**. Intuitively, we say that a surface is homogeneous if a piece of surface surrounding one point on it can be superposed on the surface around any other of its points. This situation happens with the plane, the sphere, and the cylinder but not with the torus of revolution or an ellipsoid.

Formally, we can say that a surface S is homogeneous if for any two points P and Q on the surface, there exists a rigid transformation that takes a piece of the surface surrounding P into a piece of the surface surrounding Q. In other words, all points have isometric neighborhoods.

The second concept we want to introduce is that of **isotropy**. It refers to the fact that at each point the surface shows the same aspect by looking at all directions. The plane and the sphere are isotropic but the cylinder is not. Beware that there is another important notion of isotropy (also called the stabilizer), which is different from this one and refers to the set of elements in a group of transformations of some space that leave fixed a given point.

We say that a point on a surface is an **umbilic (or navel) point** if all sectional curvatures at the point are equal. The points of a cylinder are not umbilics, but those of a sphere or of a plane are.

Exercises

1. Prove that the characteristic polynomial of a matrix is invariant under rigid transformations.

2. For each of the quadric surfaces with the same rank and signature, give metric properties, such as symmetries and extensions, which show the difference among the different types.

3. Make a drawing of a surface defined by the equation
$$3x^2 - y^2 + 4xz - 10x + 2y - 4z + 3 = 0.$$

4. Without making any computation, prove each of the following statements. (a) The curvature of a plane at each of its points is 0. (b) The curvature of a sphere takes the same value at each of its points (determine that value). (c) The curvature of a cylinder takes the same value at each of its points (determine that value). (d) The curvature of a hyperbolic paraboloid at any of its points is negative.

5. Prove Formula (1.7). (Suggestion: see [DoC].)

6. Compute the curvature of each of the following surfaces at the points indicated.
   (a) Hyperbolic paraboloid: $x^2 - y^2 = z$, at $(0,0,0)$. What is the tangent plane at the origin?
   (b) Quartic of revolution: $z = x^4 + 2x^2y^2 + y^4$, at $(0,0,0)$. What is the tangent plane at the origin?
   (c) Torus of revolution: $x^4 + y^4 + z^4 + 2x^2y^2 + 2x^2z^2 + 2y^2z^2 - 10x^2 - 10y^2 + 6z^2 + 9 = 0$, at the points $(0,1,0)$, $(0,2,1)$ and $(0,3,0)$.

1.4 Cylinders and tori

In this section we use a concept which is fundamental in all branches of mathematics, that of **equivalence class**, and we introduce the concept of **geodesic**: a trajectory on a surface that minimizes, at least locally, the distance between two points of the surface.

We use equivalence relations (see Appendix 5.2) to construct new geometrical objects departing from other known objects, so that the new objects preserve important properties of the original ones.

For instance, we shall construct a cylinder and a **plane torus** which inherits a metric, a way of measuring, from that in the plane. Therefore these are surfaces of constant Gaussian curvature equal to zero. This cylinder and this torus are geometrically distinct from the cylinder and the torus of revolution contained in $\mathbb{R}^3$, from which they inherit a metric, distinct from the one induced from $\mathbb{R}^2$ as we do below. The new objects we get via equivalent relations are surfaces (or more generally manifolds, see the Appendix 5.1) defined by mathematical means, and they do not need to have a previously specified environment, though it is an exercise to show that all manifolds can be realized in some Euclidean space).

As the reader may expect, the geodesics in the plane are the straight lines. Since the metric on the cylinder and the torus that we shall construct come from the Euclidean metric in $\mathbb{R}^2$, and the distances between "nearby points" in the torus and the cylinder are just the distances between the corresponding "nearest" points in $\mathbb{R}^2$, one has that the geodesics in these surfaces are the image of the lines in $\mathbb{R}^2$ under the natural projection.

When we were children, to construct a cylinder (rather, a portion of a cylinder) we used to take a rectangle made out of pasteboard and glue two of its parallel sides.

Now let us take vertical strips of width 1 in $\mathbb{R}^2$, as shown in Figure 1.19. We say that two points **belong to the same equivalence class** (see Section 5.2) if their coordinates have the same ordinate, and their abscissas differ by an integer number, that is,

$$(x,y) \sim_1 (x',y') \quad \text{if} \quad y = y', \ x - x' \in \mathbb{Z}.$$

Then in the vertical strip

$$\bar{B} = \{(x,y)|\, 0 \le x \le 1\},$$

there is one and only one representative element of each class of equivalence, except for the points of the outer edges with the same ordinate, because they are representative elements of the same class of equivalence. Identifying those points is equivalent to gluing two parallel edges of the pasteboard; thus, we obtain an object which may be thought of as an infinite cylinder and which may be represented as

$$\mathrm{Cil} = \mathbb{R}^2/\sim_1, \quad \text{where} \quad (x,y) \sim_1 (x',y') \quad \text{if} \quad y = y', \ x - x' \in \mathbb{Z}.$$

The same cylinder may be obtained from any vertical strip of width 1 when gluing the edges, for a strip of this type contains a representative element of each class of equivalence, except for the points of the edges.

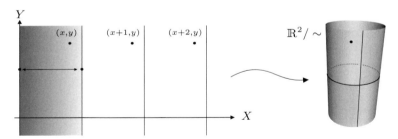

Figure 1.19: By identifying the points of $\mathbb{R}^2$ with the same ordinate and $x - x' \in Z$, we obtain a cylinder.

Actually, we call a set $\mathcal{R}$ which satisfies the following two conditions a **fundamental region**:

(i) Any point P of the plane has at least one representative element $P' \in \mathcal{R}$.

(ii) If P' is an **interior point** of $\mathcal{R}$, then P' is not equivalent to any other $P'' \in \mathcal{R}$.

1.4. Cylinders and tori

The **interior of a set** is the greatest open set contained in the set; hence, an interior point belongs to a disc (without the boundary) completely contained in the set.

One way of constructing a fundamental region is to take a point P and look for its closest equivalents, P' and P''. The perpendicular bisectors of the segments $P''P$ and PP' bound a strip in $\mathbb{R}^2$ which is a fundamental region. If we take P to be the point $(1/2, 0) \in \mathbb{R}^2$, this construction yields the above strip $\bar{B}$ as fundamental domain.

The cylinder that we get in this way shares many qualities with the plane, such as all the local metric properties, but it also has essential differences as we shall see. Among those differences, one is that any straight line parallel to the X-axis in the plane becomes a closed curve in the cylinder, which results from identifying the extremes of any piece of length 1.

A closed curve of that type in the cylinder can be distinguished from a circular line on the plane by the fact that while the latter can be **deformed continuously** into a point without leaving the plane, a "circular line" constructed as above on the cylinder, can not be deformed continuously into a point without leaving the cylinder (see Figure 1.20).

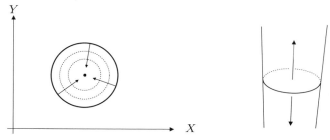

Figure 1.20: There are closed curves on a cylinder which cannot be deformed into a point of the cylinder without leaving it.

The study of closed curves in a surface leads to a very important concept in geometry, called the **fundamental group**, but its study is not in the scope of this book. The reader interested in this topic may consult [F].

Another related way of establishing this difference arises from observing that every circular line on the plane is the **boundary of a bounded region** of the plane: the interior of the disc. Instead, on the cylinder a "circular line" resulting from that identification does not determine a bounded region. This leads to another important notion in topology, the **homology groups** of spaces.

Instead, the cylinder and the plane share the **geodesics**, which are curves that minimize the distance between two points sufficiently closely, in the sense stated in the following theorem.

Theorem. *The shortest trajectory between two points P and Q on the plane is the rectilinear segment between P and Q.*

Proof. Let us remember that the length of a differentiable parametrized curve $\alpha : (c,d) \to \mathbb{R}^2$ from $\alpha(a) = P$ to $\alpha(b) = Q$ (where $[a,b] \subset (c,d)$), is

$$l(\alpha) = \int_a^b ||\alpha'(t)||dt.$$

If $\widehat{v}$ is a fixed vector of norm 1, the derivative of $\alpha(t) \cdot \widehat{v}$ is $\alpha'(t) \cdot \widehat{v}$; therefore, by integrating the latter function we obtain

$$(Q - P) \cdot \widehat{v} = \int_a^b \alpha'(t) \cdot \widehat{v}\, dt \leq \int_a^b ||\alpha'(t)||dt = l(\alpha),$$

where the inequality between both integrals is a consequence of the inequality (point to point) between the functions of the integrands.

Now it suffices to take $\widehat{v} = \frac{Q-P}{||Q-P||}$ so that the first part of the previous expression becomes $||\alpha(b) - \alpha(a)||$, which proves that the length of the segment is less than or equal to the length of any other curve joining P and Q. $\square$

Thus, given any two points P and Q of the cylinder, if P' is a point of $\mathbb{R}^2$ which is mapped into P, among all the points of $\mathbb{R}^2$ which are mapped into Q there exists one, denoted by Q', which is the closest to P'. The rectilinear segment between P' and Q' determines a curve in the cylinder and if there existed a curve from P to Q on the cylinder shorter than that corresponding to the segment, that curve would generate a curve in $\mathbb{R}^2$ with smaller length than that of the segment, which is not possible.

The **plane torus** — metrically different from the torus of revolution of Exercise 6(c), Section 1.3, as seen below — has similar differences and resemblances with the plane. The plane torus is formed by the equivalence classes of points of $\mathbb{R}^2$ under the relation

$$(x,y) \sim_2 (x',y') \quad \text{if} \quad x - x' \in \mathbb{Z},\ y - y' \in \mathbb{Z},$$

that is, now two points of the plane are related if the difference of their abscissas and that of their ordinates are both integer numbers.

Just as the first condition leads us to a vertical strip as fundamental domain, the second condition leads us to a fundamental region which is a square of side 1. This square is the intersection of a vertical strip as before, with a horizontal strip. The parallel edges of the square are identified by the equivalence relation (see Figure 1.21). The reader can easily see that the resulting object is a torus $S^1 \times S^1$, although if he or she tries to construct it with a square made out of paper, he or she will confront the problem that when gluing the last two edges, the paper gets crumpled.

The plane torus can not be constructed without corrugations or folds in our three-dimensional Euclidean space, since as on the cylinder every circular line constructed by the identification has length equal to 1, the length of one side of

1.4. Cylinders and tori

the square, on the plane torus every circular line constructed by each of the two identifications has that same length. That is why the paper corrugates when trying to glue the edges of the cylinder.

Moreover, at each point of the plane torus, the Gaussian curvature vanishes, $K(P) = 0$, in contrast with the torus of revolution where there exist elliptic, parabolic, and hyperbolic points (Exercise 6(c) of Section 1.3).

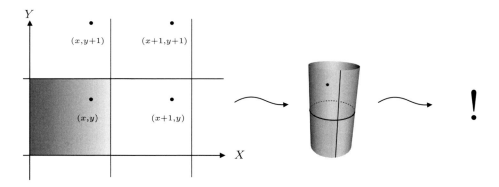

Figure 1.21: When identifying the points of $\mathbb{R}^2$ with $x - x' \in \mathbb{Z}$ and $y - y' \in \mathbb{Z}$, we obtain a plane torus (which does not fit in $\mathbb{R}^3$).

In this case, there exist two classes of closed curves which can not be deformed into a point, neither can they be deformed into each other: one class for each pair of parallel sides of the square.

Although these results establish an essential difference between the Euclidean plane and the plane torus, from the local metric viewpoint they look the same: when determining which are the geodesics in the torus, it suffices to consider that for any two different points P and Q on the torus, if $P' \in \mathbb{R}^2$ is mapped into P when taking equivalence classes, there exists a point $Q^* \in \mathbb{R}^2$ which is mapped into Q and is the closest to P' among the nine possible points (see Figure 1.22).

The rectilinear segment from P' to Q^* is mapped into a curve of the torus which necessarily minimizes the length of the path from P to Q on the plane torus. For if there existed a curve on the plane torus from P to Q of length less than that of the rectilinear segment, that curve would generate a curve between P' and Q' of length less than that of the segment, which is impossible.

In other words, we consider the so-called **standard projection** of a space into the set of its equivalence classes. In this case,

$$\Pi : \mathbb{R}^2 \to Torus,$$

which maps each point (x, y) into its class $[(x, y)]$ according to the equivalence relation $\sim_2$.

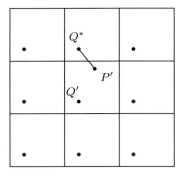 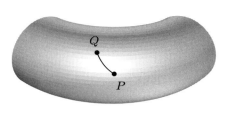

Figure 1.22: The rectilinear segment from P' to Q^* generates a curve on the torus which minimizes the path from P to Q.

The geodesics of the plane torus are the images under Π of the geodesics of $\mathbb{R}^2$, for the way of measuring in the plane torus is inherited from that in $\mathbb{R}^2$.

It is also convenient to note that the cylinder and the plane torus have a well-defined tangent plane at each of their points. This is so because when gluing the edges of a strip which is a fundamental region of the cylinder, the glued points have the same quality as all the others, particularly, curves in all directions go through them, as shown in Figure 1.23.

The tangent plane at a point P is considered formed by the velocity vectors of curves which go through P (they may be rectilinear segments centered at P). The curves can be traversed in one direction or the other and with any speed we desire; that is why the tangent vectors actually form a complete plane (see Figure 1.23).

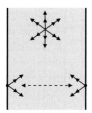

Figure 1.23: The tangent vectors at any point P of a cylinder (resp. a plane torus) form the tangent plane at P.

An interesting exercise, especially for the applications that will be studied below, is to think of a straight line drawn in $\mathbb{R}^2$ whose trace we would like to preserve when obtaining the identifications we have posed.

On the cylinder, we would obtain a "helix," and as limit cases we would have a straight line and a circle (see Figure 1.24).

1.4. Cylinders and tori

On the torus, we can obtain many types of closed curves, as occurs with the straight line through the origin whose slope is 1, since the four vertices of the square are finally identified as just one, or with any other straight line joining two points with integer coordinates, for the same reason of the previous case (see Figure 1.24).

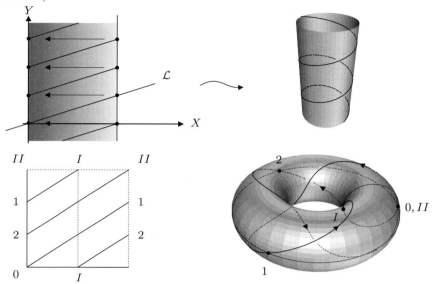

Figure 1.24: A straight line in $\mathbb{R}^2$ originates helices on the cylinder and curves which wind round the torus, sometimes densely, without self-intersections.

Furthermore, take a line through the origin in $\mathbb{R}^2$ with any slope you want which is of the form m/n with m, n positive integers. We leave as an exercise for the reader to show that this line projects to a closed curve in the torus, which winds around the torus several times before closing. Moreover, we can take all the lines in the plane with that same slope. They obviously fill out the whole plane, and therefore their images in the torus fill it out entirely, thus giving a partition of the torus into smooth curves, all of them "parallel". This is a beautiful example of an important geometric structure on surfaces (and manifolds in general) called a **foliation**.

Now take a line on the plane through the origin whose slope is irrational. The curve that is generated on the torus does not close and, furthermore, it winds around on the torus in such a way that for any point of the torus, there are points of the curve as close as we want: the curve is a **dense set** on the torus.

To prove the latter result, let us consider the straight line $y = mx$ with $m \in \mathbb{R} \setminus \mathbb{Q}$. Hence, the points on the left side of the square that are equivalent to points of the straight line of the form (n, mn), with n integer, are of the form $(0, mn - [mn])$, where $[mn]$ is the integer part of mn; we shall write $y_n = mn - [mn]$.

We divide the left side of the square in k parts, with $k \in \mathbb{N}$. Among the points $(0, y_j)$ originated by the $k+1$ points $(1, m), (2, 2m), \ldots, (k+1, (k+1)m)$, by the "principle of nested intervals", there are two of them on the same rectilinear segment of length $1/k$, that is, for some $r, s \in \{1, 2, \ldots, k+1\}$, we have

$$y_s - y_r < \frac{1}{k}.$$

The rectilinear segment between the points (r, mr) and (s, ms) is congruent with the rectilinear segment between $(0, 0)$ and $(s-r, (s-r)m)$, which means that there exist also two points of the curve on the first rectilinear segment of length $1/k$ on the left side and, accordingly, on any other of the parts in which we had divided the left side.

Since k was arbitrary, we conclude that there are points arising from the curve as close as we want to any point of the left side. Therefore, there exists a dense set of intersections of the curve with the left side. The same occurs with any other vertical segment, and that is why the curve of the plane torus originated by a straight line of irrational slope is dense on the torus and has no self-intersections.

If we now consider all the lines in the plane with that same irrational slope, we get that their images in the torus fill it entirely by parallel curves which are pairwise disjoint, but each of them is dense in the torus. This is another example of a foliation on the torus.

It is worthwhile mentioning that there is an analogous phenomenon for the curves called **hypocycloid** and **epicycloid**, described by a fixed point of a circle of radius r which rotates without slipping inside (or outside) another circle of radius R (see Figure 1.25).

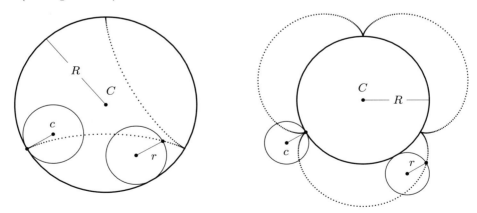

Figure 1.25: A hypocycloid and an epicycloid.

When the ratio of the radii is rational, the curve closes, but if that ratio is irrational, the point touches the circle of radius R at points that form a dense set in this circle. The proof is left as an exercise in this section.

1.4. Cylinders and tori

Up to now, we have not yet justified the title of this section, but the reader will agree that the translations $T_{(n,0)}$, corresponding to vectors of the form $(n,0)$ with $n \in \mathbb{Z}$, form a group under composition; thus, it is called a **subgroup** to the group of translations. We shall denote it as $\mathcal{T}_{\mathbb{Z}}$.

If $P_0(x_0, y_0) \in \mathbb{R}^2$ is translated by means of each element of this subgroup, we obtain points of the form $(x_0 + n, y_0)$, which we recognize as points related to P_0 according to the treatment presented at the beginning of this section where we used equivalence classes.

From this viewpoint, we say that the **orbit** of P_0 under the group $\mathcal{T}_{\mathbb{Z}}$ is

$$\mathcal{O}(P_0) = \{(x,y) \mid (x,y) = T_{(n,0)}(x_0, y_0) = (x_0 + n, y_0) \text{ for some } n \in \mathbb{Z}\}.$$

Hence, the cylinder can be seen as the **space of orbits of the points of the plane under the action of the group** $\mathcal{T}_{\mathbb{Z}}$, i.e., the space obtained by thinking of each orbit as being a "point".

Analogously, the plane torus can be seen as the space of orbits of the points of the plane under the action of the group $\mathcal{T}_{\mathbb{Z} \times \mathbb{Z}}$, whose elements are the translations $T_{(n,m)}$ associated to vectors (n,m) with integer entries.

Actually, these are two different ways of reading the same geometrical fact: (i) obtaining surfaces with a Euclidean structure departing from $\mathbb{R}^2$, whether as **spaces of identification** under an equivalence relation, which is the topological viewpoint, or (ii) as spaces of orbits of the points of $\mathbb{R}^2$, under the **action of a group**, which is the dynamical (and also algebraic) viewpoint.

The space of orbits of a group action can have very different properties. For instance, if the group acting in $\mathbb{R}^2$ is $\mathcal{T}_{\mathbb{R}}$ where $T_r \in \mathcal{T}_{\mathbb{R}}$ corresponds to the horizontal translation by a vector $(r,0)$ with $r \in \mathbb{R}$, we see that the orbit of a point $P(0, y_0)$ is the complete horizontal line $y = y_0$; accordingly, the space of the orbits of the points of $\mathbb{R}^2$ under the action of the group $\mathcal{T}_{\mathbb{R}}$ is only a straight line, which can be thought of as the Y axis. Notice that this action descends to an action of $\mathbb{R}$ on the torus, with space of orbits a circle. Of course we may define similar actions of $\mathbb{R}$ on $\mathbb{R}^2$ and on the torus, but now taking as generator a translation in the direction of a line with non-zero slope. We leave it to the reader to find out what the corresponding spaces of orbits are. Notice that in the case of irrational slope the corresponding curves in the torus are dense, as we know already. Hence the orbit space is an example of a rather "bizarre" topological space.

Exercises

1. Give four examples of homogeneous surfaces and analyze which of them contain umbilics.

2. Establish the parametric equations of the hypocycloid and prove that if the ratio of the radii is not rational, the curve touches the circle at a dense set of points.

3. If $m = p/q$, with p and q relatively prime integers, the curve on the plane torus resulting from the projection of the straight line through the origin with slope m closes when it passes again through a point with integer coordinates. Determine how many times the curve should have wound around the plane torus following the direction of the "meridians" as well as following the direction of the "parallels" so that it closes. (We call the closed curves resulting from the identification on the vertical lines meridians, and the closed curves resulting from the identification on the horizontal lines parallels.)

4. On the torus of revolution, draw the curve that corresponds to $m = 3/2$ on the plane torus.

5. What is the space of the orbits of the points of $\mathbb{R}^2$ under the action of the group of translations $\mathcal{T}_{\mathbb{R}^2}$, that is, if any point can be mapped into any other point?

1.5 Finite subgroups of $E(2)$ and $E(3)$

In this section we study some subgroups of $O(2, \mathbb{R})$ and $O(3, \mathbb{R})$ from a different viewpoint to that of Section 1.4; now we are interested in studying objects that are preserved under the action of those subgroups, that is, under any transformation of the subgroup being studied.

In Section 1.2 we already saw that the rotations in $\mathbb{R}^2$ with center at the origin form the group $SO(2)$, and the compositions of reflections on lines through the origin give rise to the group $O(2, \mathbb{R})$. $O(2, \mathbb{R})$.

Actually, the function

$$\det : O(2, \mathbb{R}) \to \mathbb{R},$$

which associates to each matrix its determinant, splits up $O(2, \mathbb{R})$ into two disjoint classes: the class of rotations and the class of reflections. When we multiply each rotation by a fixed reflection, we obtain all the reflections. Although in this book we do not use the concept of **normal subgroup**, it is worthwhile noting that the subgroup $SO(2, \mathbb{R})$ is a normal subgroup of $O(2, \mathbb{R})$ because it is the inverse image of 1 (see [B-ML]).

An object that remains invariant under any rotation about the origin is a circle with center at the origin and arbitrary radius, $\mathcal{C}(\bar{O}, r)$. A circle also remains invariant under any reflection with respect to any of its diameters, each contained in a straight line through the origin of $\mathbb{R}^2$. Thus, a circle remains invariant under every element of $O(2, \mathbb{R})$.

Let us now consider a subgroup H of $SO(2)$ which does not contain all the rotations with center at the origin. If in the subgroup there is a non-trivial rotation, by an angle θ, Ro_θ, then it must contain all of its **iterations**, that is, the composition of Ro_θ with itself any number of times it may be necessary, as well as their inverses.

1.5. Finite subgroups of $E(2)$ and $E(3)$

Exercise 2 of Section 1.4 shows that the simplest case occurs when θ is of the form $2\pi/k$ with $k \in \mathbb{Z}$. Thus, if a point P belongs to a figure $\mathcal{F}$ which must remain invariant under a subgroup containing Ro_θ, then its images under the iterations of the rotation, $Ro_\theta^{(i)}(P) = Ro_\theta \circ \cdots \circ Ro_\theta(P)$, must also belong to $\mathcal{F}$.

This is the case, for instance, of a regular polygon of n vertices inscribed in a circle with center at the origin; in Figure 1.26 we show the case of a square and that of a regular pentagon, both inscribed in $\mathcal{C}(\bar{O}, 1)$.

The square remains invariant under the rotation by $\theta = 2\pi/4 = \pi/2$ and its multiples 2θ, 3θ and $4\theta = \text{id}$; the pentagon remains invariant under the rotation by $\phi = 2\pi/5$ and its multiples 2ϕ, 3ϕ, 4ϕ and $5\phi = \text{id}$. In both cases, the number of rotations is the same as the number of sides of the polygon.

However, we can also notice immediately that the square also remains invariant under the reflections with respect to the straight lines through the midpoints of opposite sides (M and M', N and N' in the square of Figure 1.26), which are the axes in the case of the square shown in Figure 1.26, and under the reflections with respect to the diagonals (which are the straight lines joining opposite vertices). There are four reflections.

Instead, the reflections leaving invariant the pentagon correspond to straight lines containing one vertex and the midpoint of the opposite side, like A and M_A in the pentagon shown in Figure 1.26. There are five such reflections.

We call the subgroup of $O(2, \mathbb{R})$ containing both $Ro_{\pi/2}$ and $Re_{\pi/4}$ and all the products that may be formed with these transformations the **group of symmetries of the square**.

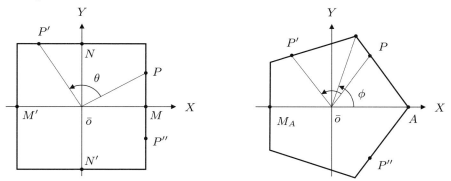

Figure 1.26: Symmetries of regular polygons.

Since the product of two rotations is a rotation and the product of one rotation with one reflection is a reflection, the four possible powers of $a = Ro_{\pi/2}$ yield the four rotations mentioned above, and when multiplied by $b = Re_{\pi/4}$, we obtain the four reflections we observe on the square.

Hence, to describe this group, two **generators** are enough: $a = Ro_{\pi/2}$ and $b = Re_{\pi/4}$. These generators satisfy the relations $a^4 = \text{id}$, $b^2 = \text{id}$ and $ab = ba^3$

(see [B-ML]).

The **group of symmetries of the pentagon** must contain $Ro_{2\pi/5}$, $Re_{2\pi/5}$, and all the possible products that may be formed with these two transformations. This group is generated by $a = Ro_{2\pi/5}$ and $b = Re_{2\pi/5}$, which satisfy the relations $a^5 = \text{id}$, $b^2 = \text{id}$ and $ab = ba^4$.

In general, the group of symmetries of a regular polygon with n sides is the so-called **dihedral group**, which consists of $2n$ elements generated by one rotation a and one reflection b which satisfy the relations $a^n = \text{id}$, $b^2 = \text{id}$ and $ab = ba^{n-1}$.

This group can be described geometrically as follows: let us consider the XY plane in $\mathbb{R}^3$ and a regular polygon with n sides contained in that plane, whose vertices belong to the unit sphere S^2.

The subgroup of $SO(3)$ formed by the rotations of XY that preserve the polygon is, clearly, a group with n elements. Actually, it is a **cyclic group** (i.e., it consists of the powers of only one element) Γ of order n, generated by the matrix

$$a = \begin{pmatrix} \cos 2\pi/n & -\sin 2\pi/n & 0 \\ \sin 2\pi/n & \cos 2\pi/n & 0 \\ 0 & 0 & 1 \end{pmatrix},$$

which clearly satisfies $a^n = \text{id}$. We have thus already n symmetries of the polygon.

The reflection of the XY plane missing to double that number can be seen as the rotation b of the sphere S^2 by an angle of 180° about one of the axes of symmetry of the polygon. That is, with respect to a straight line joining two opposite vertices if n is even, or with respect to one vertex and the midpoint of the opposite side if n is odd. The composition of b with the n elements of Γ, including the identity element, gives the other n elements of the dihedral group. It is obvious that $b^2 = \text{id}$. We leave as an exercise for the reader to verify that a and b satisfy the relation $ab = ba^{n-1}$.

It is of the utmost importance to notice that there are subgroups of the orthogonal group $O(2, \mathbb{R})$ which originate naturally when studying objects as simple as the regular polygons.

On the other hand, we can ask what is the space of orbits of the points $P \in \mathbb{R}^2$ corresponding to the group G generated by one rotation of the type $Ro_{2\pi/n}$, with $n \in \mathbb{Z}$.

It is clear that any sector of the plane bounded by two rays departing from the origin and forming an angle of $2\pi/n$ contains a representative element of each orbit, except through the points on the edges that, as in the case of the cylinder, must be identified. In this case, a point P in a boundary ray is identified with the point P' in the other boundary ray if both of them are at the same distance from the origin; in this way an infinite cone is formed, although the generatrices are not complete straight lines but only rays through the origin. That is the space of orbits of the points of $\mathbb{R}^2$ under the action of the group G. Note that the vectors tangent to the cone at the vertex do not form a plane.

We leave as an exercise to investigate what happens when we start with a rotation of $\mathbb{R}^2$ by an angle of the form $2\pi/\alpha$ with α irrational.

1.5. Finite subgroups of $E(2)$ and $E(3)$

Let us now consider the case of $O(3, \mathbb{R})$. Again, the function determinant splits up the group into two classes: the class of matrices with determinant 1 and the class of matrices with determinant -1.

Again, we call the elements of the first class, which form the subgroup $SO(3)$, rotations, whereas we call the elements of the second class reflections, and these **do not** form a subgroup; they can be obtained by multiplying each rotation by a fixed reflection.

In this case, it is also trivially true that a sphere with center at the origin and arbitrary radius remains invariant under any element of $O(3)$.

However, if we want to discover any object that remains invariant under the elements of a subgroup different from the identity element $G < O(3)$, the object must have a certain regularity, for otherwise its group of isometries will be very poor, as occurs with a cloud, whose form can be completely arbitrary. Furthermore, the more regular the object is, the greater the group G of isometries which leave it invariant.

We work with bounded objects, that is, objects that can be confined to a sphere although its radius may be very large. We call the points V in the object whose distance to the origin is a maximum **vertices**.

If the norm of V is r, V belongs to the sphere with the center at the origin and radius r as well as all its transformations under elements of G. For reasons that will be apparent later, we require the number of vertices to be finite, and each vertex can be transformed into another by means of some isometry.

Since each element of $O(3, \mathbb{R})$ is an isometry, the distance between any two vertices and its transformed vertices must be the same. The rectilinear segments determined by one vertex V_0 and any of its transformed vertices closest to V_0 are called **edges**.

For a rotation to appear in the subgroup whose axis goes through a vertex V_0, we must require that all the edges intersecting on it have the same length, and the angles between any two of them must be congruent. Since the vertices are transformed one into another, the previous considerations are valid for any vertex.

Given that the number of vertices is finite, so is the number of edges. If we take two consecutive edges from those intersecting at one vertex V_0, a_1 and a_2, when considering the other extremes of these edges, V_1 and V_2, we can take edges consecutive to a_1 and a_2, a'_1 and a'_2, the closest possible to the other: a'_1 close to a_2 and a'_2 close to a_1.

In this way we form a regular polygon, which is inscribable in a circle and therefore is planar. These polygons, all of them congruent one to another, are the **faces** of a **regular polyhedron**, and the body they delimit is a **platonic solid**. These solids were studied by the ancient Greeks and played an important role in their philosophy.

Hence, a regular polyhedron is inscribable in a sphere, and its faces are regular polygons, all of them congruent one to another. An obvious example is a cube, inscribable in the sphere whose center is that of the cube and whose diameter has the length of a main diagonal (see Figure 1.27). The other examples are the other "geometrical bodies" which we constructed in elementary school: the tetrahedron (four faces), the octahedron (eight faces), the dodecahedron (twelve faces), and the icosahedron (twenty faces).

Before we study these five geometrical objects which surprised the Greeks and, far from being just abstractions, appear repeatedly in nature, it is convenient to prove that they are the only possible regular polyhedra in three dimensions.

For that purpose, we observe that at each vertex, at least three faces must meet (on the contrary, we could not glue the faces). Thus, the number k of faces coinciding in one vertex is such that $k \geq 3$. As the number n of sides of a regular polygon is also at least 3, we have $n \geq 3$, with these results we have found lower bounds for both k and n. To find their upper bounds, it suffices to observe that the sum of the k interior angles of the faces intersecting at one vertex must be less than 2π :
$$\frac{k(n-2)\pi}{n} < 2\pi,$$
that is, $k(n-2) - 2n < 0$. By adding 4 to both members of this inequality, we obtain the inequality $(k-2)(n-2) < 4$, from which it follows that the only admissible values for a pair (k, n) are

$$(3,3),\ (3,4),\ (3,5),\ (4,3),\ (5,3),$$

each of which holds just in one of the platonic solids shown in Figure 1.27.

In Table 1.1, we show the number of vertices, edges, and faces of the platonic solids. From these data, we can observe two very interesting properties.

Solid	# vertices	# edges	# faces
Tetrahedron	4	6	4
Hexaedron (cube)	8	12	6
Octahedron	6	12	8
Dodecahedron	20	30	12
Icosahedron	12	30	20

Table 1.1: Number of vertices, edges, and faces of the platonic solids

To begin with, the numbers of the cube (or hexahedron) and those of the octahedron correspond, except that the numbers of vertices and of faces are interchanged; we observe the same behavior in the numbers corresponding to the dodecahedron and the icosahedron. This is so because when taking the barycenters of the faces as vertices and forming regular polyhedra following the same rules we used before, we obtain the other solid of the pair, including the case of the tetrahedron which must be considered as its own pair.

We say that the members of a pair are one the **dual** of the other, for reasons we shall explain below.

1.5. Finite subgroups of $E(2)$ and $E(3)$

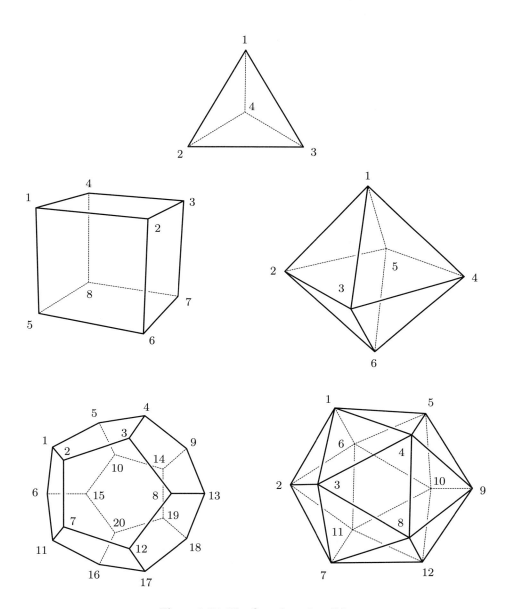

Figure 1.27: The five platonic solids

The other relevant property is that when forming, in each case, the number

$$\mathcal{X} = V - E + +F$$

where V is the number of vertices, E is the number of edges, and F is the number of faces, in every case we obtain $\mathcal{X} = 2$.

The first one to observe this property was Leonhard Euler (1707–1783); thus, this number $\mathcal{X}$ is called after him the **Euler characteristic** of the surface under study.

Of course, there exist surfaces for which the value of the Euler characteristic is not 2, as in the case of the torus. It is known that this number depends only on the surface in question and, for orientable surfaces (such as the sphere, the torus, etc.) it is given by the formula $\mathcal{X} = 2 - 2g$ where g is the genus of the surface. Notice that the platonic solids are all homeomorphic to the 2-sphere.

If we split up the torus into triangles, as shown in Figure 1.28 (notice that the triangles intersect in one side, one point, or they do not intersect at all), by counting the faces, edges, and vertices we can verify that $\mathcal{X} = 0$:

- The total of vertices is 8, 4 "inside" and 4 "outside".

- There are 4 edges inside and 4 outside, plus 4 more edges at each interior vertex (each of these edges intersects with an exterior vertex); the total of edges is $4 + 4 + 16 = 24$.

- At each interior vertex, 6 faces meet, but 4 of them also intersect with another exterior vertex; thus, we shall count only 2 of those. Hence, the result is $4 \times 4 = 16$ faces.

Accordingly, the Euler characteristic of the torus is

$$\mathcal{X} = 8 - 24 + 16 = 0.$$

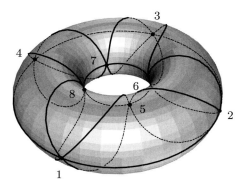

Figure 1.28: Euler's characteristic is 0 for a torus.

1.5. Finite subgroups of $E(2)$ and $E(3)$

The Euler characteristic is a concept that nowadays belongs to the area of topology. It can be proved that it does not depend on the "triangulation", whenever the triangles intersect as was described above; that is why it is a characteristic of a "topological object" (cf. [DoC]).

Let us see what the groups of symmetries of the platonic solids are. We leave the case of the tetrahedron to the reader (see Exercise 5 of this section). However, we expect that the analysis of the cases of the cube and the dodecahedron will allow him or her to prove that the symmetries of the tetrahedron form a group isomorphic to the group of permutations of four symbols (which correspond to the vertices). This group is usually denoted as S_4 and is called the **symmetric group** on four symbols (see Appendix 5.2).

As we did in Section 1.1, we place the cube so that its faces are parallel to the coordinate planes and its center is the origin. Thus, the vertices belong to a sphere with center at the origin, forming the symmetries of the cube as a subgroup of $O(3, \mathbb{R})$ (see Figure 1.29, which is analogous to Figure 1.12).

The task now is to determine the subgroup of $O(3, \mathbb{R})$ that leaves the cube invariant. According to what we have seen, it suffices to restrict our work to the subgroup of the rotations, $SO(3, \mathbb{R})$, since when composing any of them with the antipodal map, given by the matrix

$$\begin{pmatrix} -1 & 0 & 0 \\ 0 & -1 & 0 \\ 0 & 0 & -1 \end{pmatrix},$$

which evidently fixes the cube by applying each vertex into its diametrically opposite vertex, we obtain all the reflections that leave the cube invariant, as we have already explained above.

In contrast with the case of the tetrahedron, we can not think of permuting arbitrarily the 8 vertices, since in the cube there are vertices whose distance from one to another is l — the length of one edge — while others are $\sqrt{2}l$ apart — the length of a diagonal of one face — and others are at distance $\sqrt{3}l$ — the length of one diagonal of the cube.

The task becomes easier if we note that the diagonals of the cube, which are 4 (shown in Figure 1.29),

$$D_1 = AT; \quad D_2 = BU; \quad D_3 = CR; \quad D_4 = DS,$$

must be mapped one into another under any rotation that leaves the cube invariant, that is, *any rotation that fixes the cube produces a permutation of four objects*, the diagonals.

To prove that any permutation of D_1, D_2, D_3, and D_4, arises from a rotation of the cube, we take two diagonals and exhibit a rotation that exchanges them, leaving the other two fixed. In this way we see that each transposition of two diagonals determines a rotation of $\mathbb{R}^3$ that preserves the cube.

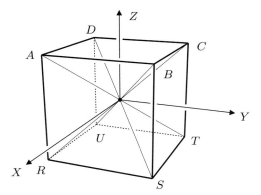

Figure 1.29: The rotations that fix the cube take diagonals into diagonals.

Hence, since transpositions generate all the permutations and the composition of two rotations is another rotation, any permutation of four objects can be obtained as the image of a rotation of the cube around one of its diagonals.

The rotation R that interchanges D_1 and D_2 and leaves fixed the other two diagonals has as an axis the straight line through the midpoints of the opposite sides AB and TU, being the rotation about that axis of an angle of π.

To verify that $R(D_1) = D_2$ and $R(D_2) = D_1$, it suffices to observe that $A \mapsto B$, $U \mapsto T$, $B \mapsto A$ and $T \mapsto U$. The other two diagonals are mapped each into itself because $C \mapsto R$ and $R \mapsto C$, which is D_3 to D_3 and, since $D \mapsto S$ and $S \mapsto D$, D_4 also transforms into itself. In the language of permutations, R corresponds to the transposition (12), and there are $6 = 4 \cdot 3/2$ transpositions of 4 objects.

As an exercise, let us see what is the resulting rotation of composing two rotations obtained by transpositions.

Let us call R' the rotation corresponding to the transposition (23); the composition of R' with R corresponds to first applying the transposition (12) and then (23), and this takes 1 to 3 and 3 to 2:

$$(23)(12) = (132).$$

Since the diagonal D_4 remains fixed, the rotation corresponding to the permutation (132) has D_4 as axis and the angle of around that axis is a multiple of $2\pi/3$, for there are 3 edges that concur at the vertex D.

If we prove that there are as many rotations as elements in the group S_4, that is, 24, then we shall have proved that the correspondence between the permutations of the diagonals and the rotations of the cube is an isomorphism of the symmetric group in four letters, S_4, and the subgroup of $SO(3)$ which leaves the cube invariant.

1.5. Finite subgroups of $E(2)$ and $E(3)$

With that, we shall have proved as well that the group of isometries of the tetrahedron is a subgroup of the group of isometries of the cube. That follows immediately from the fact that, in the cube, there are two inscribed tetrahedra (one reflected from the other): those formed with the *diagonals of the faces* that intersect with one vertex (see Exercises 5 and 7).

To count the rotations that preserve the cube, let us analyze which are the axes of rotation and the possible angles of rotation that do not produce the identity:

- If the axis is the straight line joining the midpoints of two opposite sides, the angle of rotation can only be π; since there are 6 pairs of opposite edges, we have 6 rotations of this type.

- If the axis is the straight line joining the centers of opposite faces, the angles of rotation that leave the cube fixed are multiples of $\pi/2$. There are 3 of these axes and they do not produce the identity. Since there are 3 pairs of opposite faces, we have 9 rotations of this type.

- If the axis is the straight line joining opposite vertices, since at a vertex 3 sides concur, the possible angles of rotation are multiples of $2\pi/3$, and only 2 of them do not produce the identity. Since there are 4 pairs of opposite vertices, we have 8 rotations of this type.

If to these 23 rotations we add the identity, we have 24 different rotations, the same number as that of elements in S_4.

A good exercise for the reader is to find the rotation corresponding to the product of 2 disjoint transpositions, such as (12)(34), and to a permutation of length 4, such as (1234).

We suggest that the reader make a complete table with the correspondence among the products of transpositions and rotations (we leave it as an exercise).

Moreover, of course, we should recall that the group of all isometries of the cube consists of the 24 rotations listed above and of the 24 reflections obtained by composing each of those rotations with the antipodal transformation.

The case of the octahedron is translated into that of the cube because of duality. Moreover, the case of the dodecahedron, which is translated into that of the icosahedron for the same reason, is just a little bit more complicated than that of the cube, as we see below.

Again not all the vertices of the dodecahedron are the same distance apart from each other; therefore, we can not arbitrarily permute its 20 vertices.

However, as in the cube, the 4 diagonals necessarily are mapped one into another under any isometry that leaves the cube invariant; in the dodecahedron there are 5 inscribed cubes which are necessarily interchanged under any isometry that leaves the dodecahedron fixed. They are the cubes whose edges are diagonals of the pentagons, one on each face (see Figure 1.30).

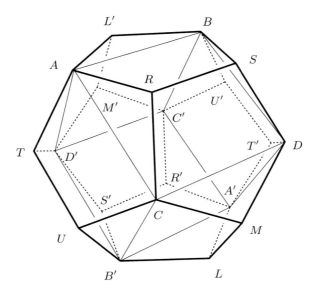

Figure 1.30: Isometries of the dodecahedron.

If we take the diagonal AB, the antipodal vertices A' and B' determine a diagonal parallel to the former; the diagonal CD, also parallel to AB (because both are parallel to the edge RS of the dodecahedron) with the antipodal vertices produces the diagonal $C'D'$. They are four parallel edges of the cube.

Another set of parallel diagonals results from considering the diagonals AC, $A'C'$, DB and $D'B'$, all of them parallel to the edge TU of the dodecahedron, and the last set of parallel diagonals is formed by AD', $A'D$, CB' and $C'B$, all of them parallel to the edge LM of the dodecahedron.

The regularity of the dodecahedron implies that the parallelepiped whose vertices are A, B, C, D, A', B', C' and D' is actually a cube.

The 5 diagonals of each face of the dodecahedron are edges of a different cube; thus, these 5 cubes are inscribed in the dodecahedron, and it is possible to take one into any of the others by means of rotations of the following types.

- Rotations whose axis pass through the centers of two opposite faces, by an angle of $2k\pi/5$, with $1 \leq k \leq 4$; there are 6 pairs of opposite faces and four rotations different from the identity for each pair. That gives 24 rotations of this type.

- Rotations whose axis pass through two opposite vertices, by an angle of $2k\pi/3$, with $1 \leq k \leq 3$; there are 10 pairs of opposite vertices and 2 rotations different from the identity for each pair. That gives 20 rotations of this type.

- Rotations whose axis pass through the midpoints of opposite sides, by an angle of $2k\pi/2$, with $k = 1, 2$; there are 15 pairs of opposite sides and only one rotation different from the identity for each pair. Thus, there are 15 rotations of this type.

If to these 59 rotations we add the identity, we have 60 rotations, which is precisely the number of elements of A_5, the **alternate subgroup**, or of even permutations, of S_5.

The reader should prove that the correspondence between the group of rotations that leave the dodecahedron invariant and the elements of A_5 is actually an isomorphism, as well as that

No isometry of the dodecahedron produces a transposition of the 5 cubes,

for any isometry of the dodecahedron that interchanges a pair of cubes necessarily also exchanges another pair.

The complete group of isometries of the dodecahedron is obtained by composing each of the rotations with the antipodal map; therefore, it is isomorphic to $A_5 \times \{\pm 1\}$, but not to S_5, according to the remark we have just made.

The discussion above shows that the group of symmetries of the cube is a subgroup of the group of symmetries of the dodecahedron and, since the group of symmetries of the cube contains the group of symmetries of the tetrahedron as a subgroup, it turns out that

S_4 is a subgroup of the group of symmetries of each platonic solid.

Thus, we have showed the following information about the groups of symmetries of the platonic solids (the reader should have taken care of the tetrahedron):

- For the tetrahedron, this is the symmetric group S_4, with A_4 as subgroup of rotations.

- For the cube and the octahedron this is $S_4 \times \{\pm 1\}$, with S_4 as subgroup of rotations.

- For the dodecahedron and the icosahedron this is the alternating group $A_5 \times \{\pm 1\}$, with A_5 as subgroup of rotations.

This determines completely the algebraic structure of these groups of symmetries.

Exercises

1. Prove that $SO(2, \mathbb{R})$ is a group isomorphic to the multiplicative group of the complex numbers of norm 1.

2. Consider the function determinant from $GL(3, \mathbb{R})$ to $\mathbb{R} - \{0\}$. Is the image of $O(2, \mathbb{R})$ under that function a subgroup of the multiplicative group $\mathbb{R} - \{0\}$?

3. Given a reflection on the XY plane with respect to a straight line through the origin, find the matrix of the rotation about the origin of the XYZ Cartesian space whose effect on XY is the initial reflection.

4. Verify that the transformations given in the geometric description of the dihedral group of order $2n$, a and b, satisfy $ab = ba^{n-1}$.

5. Prove that the group of symmetries of the tetrahedron is isomorphic to S_4. (Suggestion: Exhibit a reflection that interchanges two vertices and leaves fixed the other two, which is equivalent to a transposition of the vertices. See Appendix 5.3.)

6. Establish explicitly the isomorphism between the group of the rotations that leave the cube invariant and the symmetric group S_4.

7. Draw the two tetrahedra inscribed in a cube, formed with the diagonals of the faces that intersect with one vertex. Justify the statement:
   The group of isometries of the tetrahedron is a subgroup of the group of isometries of the cube.

8. Determine the subgroup of $SO(3)$ that consists of all the rotations that leave the north pole of the sphere S^2 fixed. (The subgroup is called the **stabilizer** of the north pole.)

9. Prove that $SO(3)$ acts transitively in points of S^2, that is, for any two points $P, Q \in S^2$, there exists a rotation that takes P into Q.

10. Given $P, Q \in S^2$, determine how many elements of $SO(3)$ take P into Q.

11. What is the Euler characteristic of the double torus shown below?

12. Split up a closed disk into four "triangles" and compute its Euler characteristic.

13. Verify that if in one of the "triangles" of the disk an interior point is fixed, and from it straight lines are drawn to the vertices of the "triangle," the value of the Euler characteristic with this new triangulation is the same as that of the former.

1.6 Frieze patterns and tessellations

Many buildings present decorative stripes, called **frieze patterns**, along their façades or on their interior walls. The design of a frieze pattern is formed by one single figure which is repeated indefinitely, as occurs with those shown in Figure 1.31 obtained by rotating an engraved and inked cylinder.

1.6. Frieze patterns and tessellations

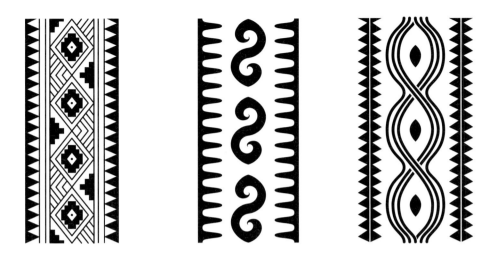

Figure 1.31: Frieze patterns of Mexican origin.

It is also common to find walls and complete floors of buildings covered by one single figure which is repeated regularly, as occurs with those shown in Figure 1.32. In colloquial language, we call the pieces used to cover the floor tiles; however, in mathematics we call the whole design formed on the floor with those pieces a **tiling or tessellation**.

The problem we study in this section consists of finding all possible types of frieze patterns and tessellations in regards to the subgroups of E(2) which preserve them. This problem was solved in practice long ago; after it was stated mathematically with precision, it inspired generalizations to elliptic and hyperbolic geometries, as we shall see in Chapters 3 and 4.

Figure 1.32: Tessellations of Arabian origin.

To create a frieze pattern, we must draw a straight line $\mathcal{L}$ following these two rules:

1. Use a pattern cut out on a piece of cardboard, as shown in Figure 1.33, with its base standing over a line $\mathcal{L}$, and denote by a the length of its base. Then repeat the pattern by translating the cardboard a length a along the line $\mathcal{L}$. You can use it on both sides of the line $\mathcal{L}$, with its base remaining on $\mathcal{L}$. Now iterate this process.

2. If the pattern is used on one side of the straight line $\mathcal{L}$, it is not permitted to use it again on the same side with another position, for that would be equivalent to having another pattern.

Hence, the **group of isometries of a frieze pattern** consists of the elements of $E(2)$ which leave a straight line fixed and contain an infinite cyclic subgroup of translations.

We shall demonstrate how to construct all the possible types of frieze patterns using the triangle that results from cutting a rectangle along one of its diagonals as a pattern (in practice, we must leave a small edge). Later on, we shall prove that actually there are no more possible types of frieze patterns.

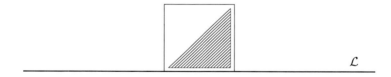

Figure 1.33: A simple pattern with no internal symmetries fixing the base.

It is convenient to denote as rA the points of $\mathcal{L}$ corresponding to segments of length ra, where r is of the form $n/2$, with n an integer. For the analysis of the isometries of the plane that leave the frieze pattern invariant, we shall begin by listing the isometries with which we generated the frieze pattern.

1º. The easiest way is to use the pattern on each segment by displacing it. The frieze pattern obtained after painting the blank space is $\mathcal{F}_1$ of Figure 1.34.

This frieze pattern remains invariant under the infinite cyclic group G_1 of isometries generated by the translation $T_{\bar{a}}$, where $\bar{a}$ is a vector parallel to $\mathcal{L}$ and of length a. The following notation is read as "G_1 is the group generated by $T_{\bar{a}}$:"

$$G_1 = <T_{\bar{a}}>.$$

2º. We can also use the pattern on both sides by rotating it, in each case, 180° about the point $(1/2)A$ of $\mathcal{L}$, and then displacing the figure we have obtained by means of $T_{\bar{a}}$. The frieze pattern $\mathcal{F}_2$ that results by painting the blank space of the pattern in each segment on both positions is the second one of Figure 1.34.

1.6. Frieze patterns and tessellations

The group G_2 of isometries that leave this frieze pattern invariant can be generated with $\sigma_{1/2}$, the rotation of 180° about the point $(1/2)A$, and $T_{\bar{a}}$. We know that $\sigma_{1/2}^2 = \text{id}$, and in the figure, it is easy to verify that $\sigma_{1/2} \circ T_{\bar{a}} \circ \sigma_{1/2} = (T_{\bar{a}})^{-1}$; thus, to introduce G_2 we also write these relations:

$$G_2 = <\sigma_{1/2}, T_{\bar{a}} |\, \sigma_{1/2}^2 = \text{id},\, \sigma_{1/2} \circ T_{\bar{a}} \circ \sigma_{1/2} = (T_{\bar{a}})^{-1} >.$$

3°. If we use the pattern on both sides of the straight line by reflecting the pattern with respect to $\mathcal{L}$ and then making a translation by T_a, the obtained frieze pattern is $\mathcal{F}_3$, the third one of Figure 1.34.

The group G_3 of isometries that leave $\mathcal{F}_3$ invariant is generated by the translation $T_{\bar{a}}$ and the reflection with respect to $\mathcal{L}$, which we denote as $R_{\mathcal{L}}$. The composition of these isometries is commutative; thus, in this case the only relation is $R_{\mathcal{L}}^2 = \text{id}$, and we write

$$G_3 = <R_{\mathcal{L}}, T_{\bar{a}} |\, R_{\mathcal{L}}^2 = \text{id} >.$$

4°. The fourth frieze pattern of Figure 1.34, $\mathcal{F}_4$, results from reflecting the tile in each case with respect to its right vertical edge.

The group G_4 of isometries that leave $\mathcal{F}_4$ invariant can be generated by the translation $T_{2\bar{a}}$, for we have formed a figure of twice the length of the original pattern, and the reflection with respect to the straight line $\mathcal{V}$ perpendicular to $\mathcal{L}$ at the point A, which we denote as $R_{\mathcal{V}}$. We know that $R_{\mathcal{V}}^2 = \text{id}$, and it is easy to prove that $R_{\mathcal{V}} \circ T_{2\bar{a}} \circ R_{\mathcal{V}} = T_{2\bar{a}}^{-1}$, thus, in this case we write

$$G_4 = <R_{\mathcal{V}}, T_{2\bar{a}} |\, R_{\mathcal{V}}^2 = \text{id},\, R_{\mathcal{V}} \circ T_{2\bar{a}} \circ R_{\mathcal{V}} = T_{2\bar{a}}^{-1} >.$$

5°. We can also form a frieze pattern $\mathcal{F}_5$, which reminds us of our footsteps on the sand, by using the isometry precisely called a **step** and which we denote as γ; a step results from composing the translation by a with the reflection with respect to $\mathcal{L}$: $\gamma = T_{\bar{a}} \circ R_{\mathcal{L}}$.

In this case it is not necessary to include a translation among the generators, for $\gamma^2 = T_{2\bar{a}}$; thus, the group G_5 is

$$G_5 = <\gamma >.$$

Note that although both groups G_1 and G_5 are cyclic and infinite, the elements of $E(2)$ generated by each of them are different isometries; that is why the frieze pattern $\mathcal{F}_1$ is not invariant under a step.

6°. A sixth frieze pattern results if we use the two reflections $R_{\mathcal{L}}$ and $R_{\mathcal{V}}$, and (necessarily) the translation $T_{2\bar{a}}$.

The frieze pattern $\mathcal{F}_6$ (see Figure 1.34) remains invariant under the group G_6 generated by these three isometries. Since G_6 contains the two generators of G_4, the following relations hold in this case, as well as that obtained from $R_{\mathcal{L}}^2 = \text{id}$ and that of the square of their composition, $(R_{\mathcal{L}} \circ R_{\mathcal{V}})^2 = \sigma_A^2 = \text{id}$. Thus, in this case we write

$$G_6 = <R_{\mathcal{V}}, R_{\mathcal{L}}, T_{2\bar{a}} |\, R_{\mathcal{V}}^2 = \text{id},\, R_{\mathcal{L}}^2 = \text{id},\, (R_{\mathcal{L}} \circ R_{\mathcal{V}})^2 = \text{id},\, R_{\mathcal{V}} \circ T_{2\bar{a}} \circ R_{\mathcal{V}} = T_{-2\bar{a}} >.$$

7º. Finally (we justify below our statements), the frieze pattern $\mathcal{F}_7$ is constructed by first reflecting the pattern with respect to the right vertical line, and then rotating it by 180° about the point of $\mathcal{L}$ denoted as $2A$ (see F1 of Figure 1.34).

The group G_7 of isometries that leave $\mathcal{F}_7$ invariant has the following representation:

$$G_7 = < R_\mathcal{V}, \sigma_{2A} \mid R_\mathcal{V}^2 = \mathrm{id}, \sigma_{2A}^2 = \mathrm{id}, (R_\mathcal{V} \circ \sigma_{2A})^2 = T_{4\bar{a}} >.$$

Note that also in this case, it is not necessary to include a translation among the generators, for $(R_\mathcal{V} \circ \sigma_{2A})^2 = T_{4\bar{a}}$.

We leave as an exercise for the reader to analyze which are the axes and centers of symmetry of the frieze patterns we have constructed (when those exist), and to determine the so-called **fundamental domains**: that is, the pieces of frieze patterns that build up the complete frieze when they are translated by the elements of the corresponding subgroup.

Remember that we are interested in proving that there are no more types of frieze patterns than those corresponding to Figure 1.34, in the sense that the obtained subgroups are all of the subgroups of isometries that leave a straight line fixed and satisfy conditions 1 and 2.

Proposition. *The subgroups of isometries of $E(2)$ that leave a straight line $\mathcal{L}$ fixed and satisfy conditions 1 and 2 are only the subgroups G_1 through G_7.*

Proof. The isometries that leave a straight line $\mathcal{L}$ on the plane invariant, and for which the conditions 1 and 2 hold, are the following:

(i) Any translation by $n\bar{a}$, where $\bar{a}$ is a vector parallel to $\mathcal{L}$ and n is an integer, denoted as $T_{n\bar{a}}$. The restriction that the multiple must be an integer is because the pattern can not be superimposed, since the figure obtained on one segment would not then correspond to the figure of the pattern.

(ii) Any rotation by 180° about a point $rA \in \mathcal{L}$ where $r = n/2$, with n an integer, denoted as σ_{rA}.

(iii) The reflection with respect to $\mathcal{L}$, denoted as $R_\mathcal{L}$.

(iv) The reflection with respect to a straight line $\mathcal{V}$ perpendicular to $\mathcal{L}$ at a point of the type nA (remember that the pattern can not be superimposed even on a portion). Notice that using such a reflection forces the translation that leaves the design invariant to correspond to a vector $2n\bar{a}$.

(v) A "step," $\gamma = T_{\bar{a}} \circ R_\mathcal{L}$, the isometry that generates the frieze pattern $\mathcal{F}_5$.

Now let us see which groups can be generated with these isometries.

1. If a subgroup contains only translations, it must be precisely G_1, for the pattern must be used once on each segment of length a and always on the same side.

1.6. Frieze patterns and tessellations

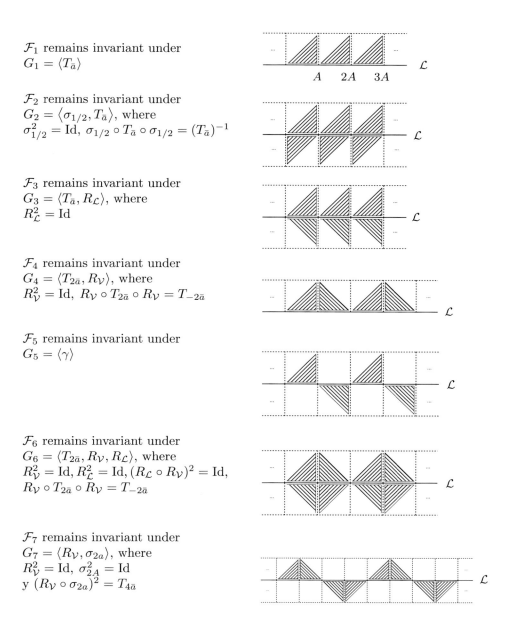

$\mathcal{F}_1$ remains invariant under
$G_1 = \langle T_{\bar{a}} \rangle$

$\mathcal{F}_2$ remains invariant under
$G_2 = \langle \sigma_{1/2}, T_{\bar{a}} \rangle$, where
$\sigma_{1/2}^2 = \mathrm{Id}$, $\sigma_{1/2} \circ T_{\bar{a}} \circ \sigma_{1/2} = (T_{\bar{a}})^{-1}$

$\mathcal{F}_3$ remains invariant under
$G_3 = \langle T_{\bar{a}}, R_{\mathcal{L}} \rangle$, where
$R_{\mathcal{L}}^2 = \mathrm{Id}$

$\mathcal{F}_4$ remains invariant under
$G_4 = \langle T_{2\bar{a}}, R_{\mathcal{V}} \rangle$, where
$R_{\mathcal{V}}^2 = \mathrm{Id}$, $R_{\mathcal{V}} \circ T_{2\bar{a}} \circ R_{\mathcal{V}} = T_{-2\bar{a}}$

$\mathcal{F}_5$ remains invariant under
$G_5 = \langle \gamma \rangle$

$\mathcal{F}_6$ remains invariant under
$G_6 = \langle T_{2\bar{a}}, R_{\mathcal{V}}, R_{\mathcal{L}} \rangle$, where
$R_{\mathcal{V}}^2 = \mathrm{Id}, R_{\mathcal{L}}^2 = \mathrm{Id}, (R_{\mathcal{L}} \circ R_{\mathcal{V}})^2 = \mathrm{Id}$,
$R_{\mathcal{V}} \circ T_{2\bar{a}} \circ R_{\mathcal{V}} = T_{-2\bar{a}}$

$\mathcal{F}_7$ remains invariant under
$G_7 = \langle R_{\mathcal{V}}, \sigma_{2a} \rangle$, where
$R_{\mathcal{V}}^2 = \mathrm{Id}$, $\sigma_{2A}^2 = \mathrm{Id}$
y $(R_{\mathcal{V}} \circ \sigma_{2a})^2 = T_{4\bar{a}}$

Figure 1.34: The different types of frieze patterns and their groups.

2. If a group contains an isometry of the type (ii), with $r = 1/2$, the pattern is used on both sides on each segment and the translation necessarily is $T_{\bar{a}}$; thus, the subgroup they generate is G_2.

3. If the subgroup contains an isometry of the type (iii), the pattern has been used on both sides and the translation must be $T_{\bar{a}}$. The generated group is G_3.

4. If the subgroup contains an isometry of type (iv), we can add another of type (i), with $n = 2$. The subgroup they generate is precisely G_4.

5. If the subgroup contains the isometry (v), it automatically contains the translation by $T_{2\bar{a}}$ and none corresponding to a vector of smaller length; thus, the group it generates is G_5.

6. If the subgroup contains an isometry of type (ii) and one of type (iii), one of type (iv) automatically appears, and since all of them are **involutions** (their square is the identity element), it is necessary to add one translation which corresponds to $2\bar{a}$; thus, the group is G_6.

7. If now we allow for an isometry of type (ii) and another of type (iv), the group they generate is G_7.

8. An isometry of type (ii) and another of type (v) which do not produce a superposition of the pattern are $\sigma_{1/2}$ and γ; however, then the straight line $\mathcal{V}$ is necessarily an axis of symmetry (prove it), and the group is isomorphic to G_7.

9. The isometries (iii) and (iv) produce, by composition, one of type (ii); all of them are involutions, and then the translation by $T_{2\bar{a}}$ must be introduced in the group. Thus, the group is G_6.

10. If we can apply the isometries (iii) and (v) to the pattern, we obtain the frieze pattern $\mathcal{F}_3$, since G_3 contains a subgroup isomorphic to G_5.

11. The isometries (iv) and (v) originate, by composition, centers of symmetry (determine them); that is, there appears an isometry of type (ii), and the subgroup is again isomorphic to G_7.

The reader must prove that it is not necessary to consider groups with more than three generators. For instance, by inspecting the frieze patterns illustrated we can discover that several of them remain invariant under steps of length different to that of γ. If the reader additionally solves Exercise 6, we can consider that the proof of the proposition is complete. □

Let us now study the tessellations. Our presentation on this theme will not be exhaustive, for the proof that there are only 17 different types of tessellations for the plane is too extensive and it is not a goal of this book. We shall only generate an example of each of the possible types; from the way we shall construct them, it will become evident that they are different.

1.6. Frieze patterns and tessellations

For the reader interested in this theme, we mention several references with different in-depth approaches, among others: [Mar], [G-S], [Sp], and [C-G]. The first reference includes a proof of this result, and the last given reference is a very attractive video.

In this case the mathematical problem consists of determining the different subgroups of $E(2)$ (non-isomorphic subgroups) whose generators leave a tessellation invariant.

Of course, if the tessellation has no regularity, the only isometry that leaves it invariant is the identity. The tessellations that are invariant under a non-trivial subgroup of $E(2)$ are constructed departing from a region which repeats itself covering the whole plane. That is, a region such that its various images by the elements of the corresponding subgroup cover the whole plane and none of them overlap (except for boundary points).

Naturally, the first thing that comes to mind is to take any of the seven frieze patterns and translate it by the elements of the group generated by $T_{\bar{a}^\perp}$, where $\bar{a}^\perp$ is perpendicular to $\bar{a}$, and has the same length (width).

That is the exercise we solved as shown in Figure 1.35, where we can immediately observe that we do not obtain seven different groups for those tessellations, but only five, for the tessellation $\mathcal{M}_4$ is obtained from tessellation $\mathcal{M}_3$ by means of a rotation by $\pi/2$; accordingly, the group that leaves invariant one of the tessellations is the conjugate of that which leaves invariant the other. In the group there are reflections and steps.

In algebra we say that a subgroup H' of a group G is the **conjugate** of another subgroup H, if there exists an element $g \in G$ such that any $h' \in H'$ can be obtained as $h' = ghg^{-1}$.

Hence, the isometries g' that leave $\mathcal{M}_4$ invariant can be obtained from the elements $g \in \mathcal{M}_3$ by means of the **conjugation** given by the rotation by $\pi/2$: $g' = Ro_{\pi/2} \circ g \circ Ro_{\pi/2}^{-1}$.

The tessellation $\mathcal{M}_1$ also remains invariant under the same group as $\mathcal{M}_3$; in this case the bijection that takes one tessellation into the other is established as follows: First, we apply a rotation by $\pi/4$ (taking as center of rotation the center of any of the squares), then the distance between the axes of the steps is reduced (instead of $2||\bar{a}||$ it will be $||(1/2)\bar{d}||$, where $\bar{d}$ is the diagonal vector contained in an axis of symmetry) and, finally, the translation in the steps is $T_{\bar{d}}$. The composition of these three transformations is not an isometry, but only a change of scale, and thus it will become evident that to each isometry that leaves $\mathcal{M}_3$ invariant, there is a corresponding isometry of the same type that leaves $\mathcal{M}_1$ invariant.

The notation associated with each $\mathcal{M}_i$ is the one used in crystallography, where these groups appear naturally.

In this case, we are not able to use the same pattern to generate tessellations corresponding to every possible type; therefore, it is worthwhile to make some simple remarks and to prove a theorem (of crystallographic restriction) which naturally classifies the 17 types of tessellations.

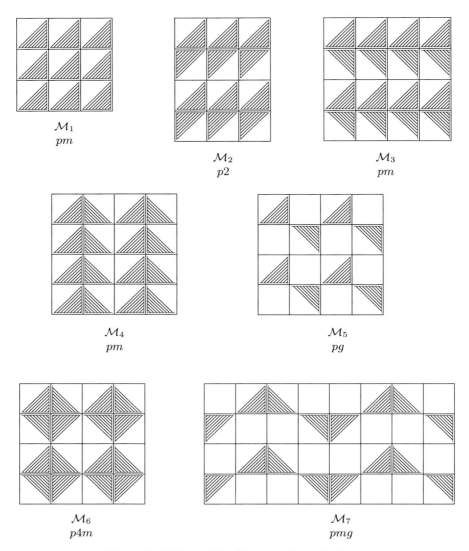

Figure 1.35: Five of the 17 types of tessellations.

1.6. Frieze patterns and tessellations

For that purpose, we first need the following definition:

A point O of the plane is an n-**center of a tessellation** if it is the center of a rotation by an angle of $2\pi/n$, with $n \in \mathbb{N}$, which leaves the tessellation invariant.

Now some useful remarks:

I. The groups of rotations that permit the invariance of a region are the cyclic groups generated by an angle of the form $2\pi/n$ with $n \in \mathbb{N}$; otherwise, the images of the region would overlap.

II. If O is an n-center of a tessellation and $O' = T(O)$, where $T \in \mathcal{G}$, being $\mathcal{G}$ the group of isometries that leave the tessellation invariant, then O' is also an n-center of a tessellation. Analogously, if $\mathcal{L}$ is the axis of a reflection that leaves the tessellation invariant, $\mathcal{L}' = T(\mathcal{L})$ is also an axis of symmetry of the tessellation.

III. If O_1 and O_2 are n-centers of a tessellation, the distance between the two of them can not be less than half of the minimum norm of the translations that leave the tessellation invariant. That is so because the composition of the rotation R_2^{-1} with center at O_2 by the angle $-2\pi/n$, and R_1 with center at O_1 by the angle $2\pi/n$ is a translation T (prove it), which must belong to the group of invariance of the tessellation, $\mathcal{G}$, and therefore is of the form $T = T_{\vec{b}}^j \circ T_{\vec{a}}^i$; accordingly,
$$R_1 = R_2 \circ T.$$
By applying both members to O_2, we obtain an isosceles triangle with vertices O_1, O_2 and $T(O_2)$. Hence, by the triangle inequality,
$$||T|| \leq 2d(O_1, O_2),$$
as we stated.

Now we can enunciate and prove the Theorem of the Crystallographic Restriction.

Theorem. *If O is an n-center of a tessellation, then n can only take the values 2, 3, 4 or 6.*

Proof. Let $Q \neq O$ be another n-center of the tessellation whose distance to O is minimum. If R is obtained by rotating O around Q by $+2\pi/n$, and S is obtained by rotating Q around R by that same angle (make a drawing), R and S are also n-centers of the tessellation and we have the equalities $d(O, Q) = d(R, Q) = d(R, S)$ and $\angle OQR = \angle QRS$.

If $S = O$, the points O, Q, and R form an equilateral triangle and $n = 6$. However, if $S \neq O$, we also have the inequality $d(O, Q) \leq d(O, S)$, because of the choice of Q. If the equality holds, we have a rhombus with equal adjacent angles, that is, a square, which implies $n = 4$. Also if $d(O, Q) < d(O, S)$, the angle must be greater than $2\pi/4$ and the only possibilities for n are 3 or 2, as we had stated. □

Hence, a natural classification of the tessellations refers to the type of n-centers it permits: a 6-center is also a 3-center and a 2-center, but there are 3-centers and 2-centers that are not 6-centers. Something analogous occurs for the 4-centers and the 2-centers, and it can be proved that the 4-centers are not compatible with the 6-centers nor with the 3-centers (see [Mar]).

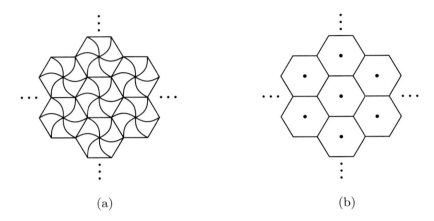

(a)   (b)

Figure 1.36: Two tessellations with 6-centers.

We begin with the study of tessellations that allow 6-centers.

Of course, we can take as a basic piece of the tessellation a regular hexagon in order to have 6-centers. We can also "mark" the faces if we wish: we shall call "mark" a curve from the center of the tile to some (or all) of the vertex, as for instance in Figure 1.36.(a); if we want our transformations to preserve these marks and the complete tesselation, then we will preserve some of its symmetries. If the hexagon has no marks, the center of any of the tessellations is a 6-center, and any vertex of one of the hexagons is a 3-center. Moreover, the straight lines containing the diagonals of the hexagon are axes of symmetry, as well as the straight lines containing the sides of the hexagon (Figure 1.36(b)). In crystallography, that type of tessellation is denoted as **p6m**.

Let us now take a basic hexagon with marks as those shown in Figure 1.36; the reflection with respect to a diagonal then does not leave the basic piece invariant, but the center of any hexagon is indeed a 6-center of the tessellation. Therefore, the tessellation (b) of Figure 1.36 remains invariant under a group different from that corresponding to the tessellation (a), for the first one does not contain reflections, and the second one does. The type of the tessellation (a) is denoted as **p6**.

If the marks of the basic hexagon are only three, as those shown in Figure 1.37(a), again the diagonals of the hexagon are not axes of symmetry, and in this case the center of any hexagon is a 3-center of the tessellation, as well as the vertices. There are no 6-centers. Their type is denoted as **p3**.

1.6. Frieze patterns and tessellations

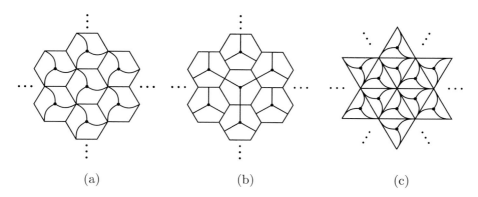

Figure 1.37: Three tessellations with only 3-centers.

However, if the three marks are apothems of the hexagon, as shown in Figure 1.37(b), and we reflect the central hexagon with respect to its sides to generate the adjacent hexagons, and we translate in the directions of the apothems by multiples of 4 times the apothem, the result is that the center of any hexagon turns out to be a 3-center, but the vertices are not. There are no 6-centers, but in this case there are axes of symmetry: the straight lines containing the apothems. The sides of the hexagons are not axes of symmetry. The corresponding notation for this type of tessellation is **p3m1**.

Finally, another tessellation with only 3-centers results from taking as a basic piece an equilateral triangle with three marks, as those shown in Figure 1.37(c), by reflecting it with respect to its sides to obtain the adjacent triangles. The center of any equilateral triangle is a 3-center, but they are not contained in axes of symmetry; however, the vertices are 3-centers contained in axes of symmetry. Their type is denoted as **p31m**.

Let us now study the tessellations with 4-centers. The basic piece is a square, with or without marks.

To eliminate reflections, we mark the square as shown in Figure 1.38(a), and to generate the adjacent tessellations we only translate. The center of any of these tessellation will be a 4-center of the tessellation, as well as the vertices, but there are no axes of symmetry. The type of this tessellation is denoted as **p4**.

If we do not mark the square and translate it by $\bar{a}$ and $\bar{a}^{\perp}$, the centers and the vertices of the squares are 4-centers, and both the sides and the diagonals of the squares are axes of symmetry of the tessellation (Figure 1.38(b)). All the 4-centers of the squares belong to axes of symmetry. The notation for this tessellation is **p4m**.

Finally, we mark the square again as in the first case, but now we reflect it with respect to its sides to generate the adjacent squares. In Figure 1.38(c) we show the resulting tessellation. The centers of the squares are 4-centers of the tessellation, but in this case the vertices are not 4-centers. However, there are axes of symmetry: the sides of the squares. The difference with tessellation (a) is that the 4-centers do not belong to axes of symmetry. The notation for this tessellation is **p4g**.

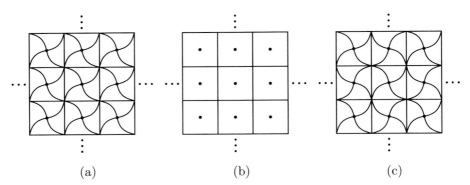

Figure 1.38: Three tessellations with 4-centers.

The tessellations that only accept 2-centers are five.

The first one is obtained by marking a square as shown in Figure 1.39(a) to break symmetries with respect to the diagonals; the adjacent squares are obtained by translating vertically and horizontally by integer multiples of a. The center and the vertices of each square are 2-centers, and there are no axes of symmetry. The crystallographic notation is **p2**.

The second tessellation is obtained by reflecting the marked square with respect to its sides; the result is shown in Figure 1.39(b). The centers of the squares are 2-centers which do not belong to axes of symmetry, but the vertices are 2-centers which do belong to axes of symmetry. The crystallographic notation for this tessellation is **cmm**.

The third tessellation of this type is obtained by reflecting the marked square with respect to its vertical sides, creating thus a frieze pattern which we then translate by the elements of the group generated by $\bar{a}^\perp$. The resulting tessellation is shown in Figure 1.39(d); the centers of the squares are 2-centers which do not belong to axes of symmetry, and these are all parallel one to another. The crystallographic notation for this tessellation is **pmg**.

A fourth tessellation uses the marked square to create steps whose translation is $T_{(1/2)\bar{a}}$; the length of the translation prevents the existence of axes of symmetry, and the centers of the squares are 2-centers. The resulting tessellation corresponds to that shown in Figure 1.39(e), whose crystallographic notation is **pgg**.

Finally, if we use as a basic piece a rectangle with no marks, and translate it horizontally and vertically by integer multiples of a, the center, the vertices and the midpoints of the sides of the rectangle are 2-centers, and all of them belong to axes of symmetry: the sides and the straight lines through the center of each rectangle which are parallel to its sides (see Figure 1.39(c)). The crystallographic

1.6. Frieze patterns and tessellations

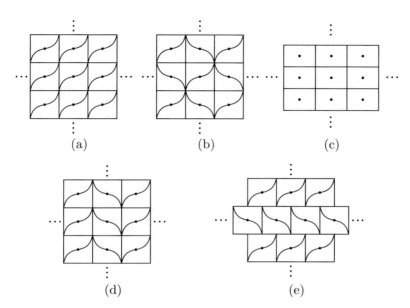

Figure 1.39: The five tessellations with only 2-centers.

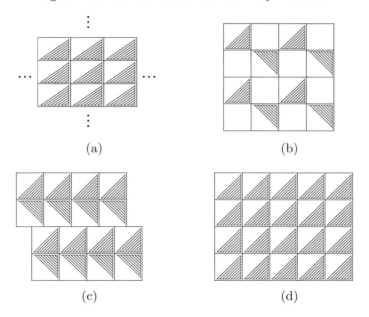

Figure 1.40: Four tessellations without centers of symmetry.

notation is **pmm**, Figure 1.39(c).

We need just four more tessellations to complete the 17 we promised; these four tessellations share the characteristic of not having centers of symmetry.

We already know two of them: they are $\mathcal{M}_1$ and $\mathcal{M}_5$ shown in Figure 1.35 (they are shown again in Figure 40(d) and (b), respectively). The first one allows axes of symmetry — the diagonals perpendicular to those that divide the two colors — and steps; their crystallographic notation is **pm**. The second one only allows steps, and their crystallographic notation is **pg**.

Another tessellation is obtained from the frieze pattern $\mathcal{F}_3$, when the additional translation is not vertical, but by integer multiples of $\bar{l} = 2\bar{a}^{\perp} + (1/2)\bar{a}$. That produces axes of steps which are not axes of symmetry, as occurs for the tessellation of Figure 1.40(c), whose crystallographic notation is **cm**.

To generate the last tessellation, we take as a basic piece a two-color rectangle, as shown in Figure 1.40(a). If we translate the rectangle horizontally and vertically by integer multiples of its base and of its height, there are no axes of symmetry. Its crystallographic notation is **p1**.

We reiterate that the proof that there are no more types of tessellations remaining invariant under a group of isometries of the plane can be found in [Mar] and [Sp].

It is worthwhile to mention that at the Alhambra, a palace in the Moorish style at Granada, Spain, there are tessellations of the 17 types exhibited in this section. The proof can be found in [Mo] (see also the videotape [C-G]).

Exercises

1. Look for illustrations of frieze patterns of ancient cultures (Greek, Celtic, Mayan, etc.) and identify under what group of transformations those frieze patterns are invariant.

2. Give an example of each type of frieze patterns constructed by using the same basic pattern (different to that used in this textbook).

3. Determine fundamental regions for each frieze pattern of Figure 33, and also for those of Exercise 1.

4. Determine the centers and axes of symmetry of each frieze pattern of Figure 33.

5. Prove that the group which leaves the following frieze patterns invariant is G_7.

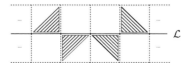 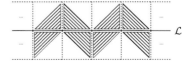

1.6. Frieze patterns and tessellations

6. Identify the group which leaves invariant the following tessellations in the Alhambra.

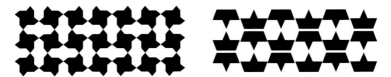

7. Design a tessellation for each of the 17 groups of isometries of the plane given in Table 1.

8. For at least three of the engravings by Escher which cover the Euclidean plane, identify the group used to construct it. (See [Cox1].)

9. Find out what is a non-periodic tessellation. (See [G-S].)

Affine geometry

2

This exposition of affine geometry is somehow different from those usually found in the literature. We have chosen a way of presenting affine geometry that constitutes a natural bridge between Euclidean geometry and projective geometry, both from the historic and the formal viewpoints. The reason this is possible is that the group of affine transformations is larger than the Euclidean group and is contained in the group of projective transformations.

We describe the relation between affine geometry and "perspective", understood as the theory developed by artists of the Renaissance who decided to create a method to take into a canvas (a flat surface), three-dimensional scenes of our environment, abandoning their plastic arts. This is presented in various texts by different authors, as for instance [Ki] and [R-S].

For instance, we **see** that the borders of a road, which actually are parallel to one another, intersect in the distance. This happens for all roads with any direction. Painters had to draw on a flat surface each point where the borders of roads meet, with different directions. They called these **vanishing points** and when they decided to place them all in a single straight line of the drawing, the so-called **horizon line**, they opened the way for mathematicians to analyze the behavior "at infinity" of many geometric objects and concepts.

For instance, when in a certain Euclidean configuration there is the possibility of having parallel straight lines, some properties must be stated in various versions, as happens with the following proposition (see Figure 2.1):

Proposition. *A quadrilateral has six vertices, except in the case of parallel sides: if only two sides are parallel, then the quadrilateral has five vertices, and if there are two pairs of parallel sides, the quadrilateral has only four vertices.*

Let us recall that a **quadrilateral** is a figure determined by four straight lines of which no three lines meet in one point, and where there is not a triple of parallel lines; a **vertex of the quadrilateral** is the intersection of two sides (see Figure 2.1).

In affine geometry, for which parallel lines intersect at a "point at infinity", any quadrilateral has six vertices, but some of these vertices are special.

Actually, the concept of parallelism has given rise to many discussions since its formulation. If we compare the statement of postulate 5 in Euclid's *Elements* — that of parallel lines — with the other four postulates (see Appendix 5.4), whose statements are brief, we can understand why some people believed one could prove postulate 5 on the basis of the other four. Euclid took care in stating it in a way

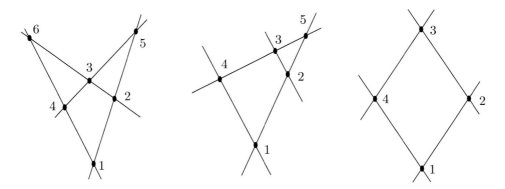

Figure 2.1: In a quadrilateral, the number of vertices depends on the number of pairs of parallel sides.

which is useful for proofs, but this made it much more complicated than the other four postulates.

To find a geometry — projective geometry — where any two "straight" lines intersect, it was necessary to understand the behavior "at infinity" of parallel lines. This took a long time: the origins of projective geometry go back to the 16th century, with Desargues and Pascal, while it was not until the 19th century when this geometry was fully formalized. There are no "special cases" in projective geometry; that is why we are able to state and prove results in a uniform way.

Affine geometry, defined by Leonhard Euler, was the step between Euclidean geometry and projective geometry. The rules established by the painters had afforded a consistent treatment of the behavior "at infinity" of parallel lines. Affine Geometry proved to be more general than Euclidean geometry, allowing the mathematical treatment of the vanishing points and the horizon line, as well as telling us how to relate mathematically two paintings, or two photographs, of the same landscape made from different positions. The reader would agree that we can recognize the landscape represented in both paintings or photographs; this means that there exist relationships in both illustrations which our brain can recognize and compare. Those relationships are the invariants of affine geometry.

This chapter is devoted to the study of affine transformations and their invariants, one of these being the type of conic. Our presentation will have "perspective" as a referent. The algebraic part is based on [B-ML]; the discovery of perspective is described in [Va].

2.1 The line at infinity

The problem a painter confronts (but not a photographer) is to represent objects of the three-dimensional space on a piece of paper or on a two-dimensional canvas.

2.1. The line at infinity

The essential part of the solution that the theory of perspective offers consists of choosing the floor as being the **Cartesian plane** and then giving the rules for sketching on the **plane of the drawing** the straight lines of the floor which are parallel. Henceforth, these two planes play different roles, and we shall devote ourselves to explaining how they are related.

We call a point of the plane of the drawing which is a meeting point of the lines corresponding to all lines on the floor parallel to a given direction a **vanishing point**; the vanishing points of each family of parallel lines of the floor form a straight line (rather a rectilinear segment) on the plane of the drawing, called the **horizon line.**

In Figure 2.2, we show a scheme of the drawing of a building located on a corner. The kerbs define two vanishing points, and these determine the horizon line. Note that the edges of the last floor of the building also meet at the vanishing points, for they are parallel to the kerbs, although they do not belong to the floor: the Cartesian planes parallel to the plane of the floor intersect at the horizon line.

The main difference between the Cartesian plane of the floor and the plane of the drawing is that on the plane of the drawing we mark points which do not exist on the Cartesian plane of the floor, the vanishing points.

These remarks give rise to two basic rules for drawing with perspective:

1. All lines parallel to the same direction in the three-dimensional space are sketched as concurring at the same vanishing point.

2. The vanishing points of the lines on the plane of the floor are sketched as belonging to the same line, the horizon line.

Figure 2.2: Sketch of a building.

These two rules suffice to make correct sketches on a piece of paper (or a canvas) of a floor of square tiles on which we may be standing: once we have sketched one of them (the shadowed quadrilateral in Figure 2.3, whose form depends on the position of the sketcher), the vanishing point of the "horizontal" sides, F_1, and the vanishing point of the "vertical" sides, F_2, determine the horizon line, which may not be horizontal on the paper. By cutting with it the straight line to which the diagonal sketched on the shadowed quadrilateral belongs, we determine a third vanishing point, F_3; this is the point where the strokes of all the diagonals of the quadrilaterals representing the other tiles meet (since on the plane of the floor they are parallel to the sketched diagonal).

To verify that the drawing of the other tiles is already determined, it suffices to sketch the straight lines passing through F_3 and each of the corners marked with H_1 and V_1. These lines include the diagonals of the right and the upper tiles, respectively, of the shadowed tile.

The intersection of $F_3 H_1$ and $V_1 F_1$ and that of $F_3 V_1$ and $F_2 H_1$ determine the points M_2 and N_2 respectively. The point M_2 is the right upper corner of the tile to the right of the shadowed tile, and the point N_2 is the right upper corner of the tile on top of the shadowed tile.

If now we draw the straight line $F_2 M_2$, its intersection with the straight line $F_1 H_1$ is the point H_2, which is the corner of the base of the second horizontal tile; drawing the line $F_1 N_2$, we cut the line $F_2 V_1$ at the point V_2, which is the corner of the left side of the second vertical tile. Thus, we have completed three tiles adjacent to the shadowed tile, since the next tile on the diagonal is also already determined.

It is clear that the process continues so that the drawing of all the tiles actually corresponds to what our eyes see.

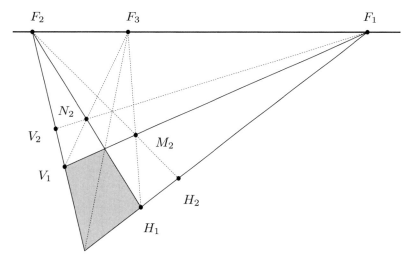

Figure 2.3: The correct drawing of a floor of tiles.

Thus, although the points corresponding to the edges of the tiles are not located at equal distances on the paper, they are completely determined by the way we sketched the first tile (it can be any quadrilateral, depending on the position of the sketcher). In other words, those points are **not** arbitrary. For example, the center of a tile is the point of intersection of the diagonals; hence, the point of the

2.1. The line at infinity

drawing that corresponds to the center of a tile is the point where the diagonals of the quadrilateral representing that tile intersect.

To formalize these ideas, scientists had to solve a problem that does not ever arise when making drawings: Since when rotating 180°, the borders of a road seem to intersect as well in the other direction, should we add one or two points "at infinity" to a straight line?

Gerard Desargues (1591–1661, French engineer) and Johan Kepler (1571–1630, German astronomer) determined that for each line there must exist only one point at infinity. This means that the topological appearance of an affine line is that of a closed curve, such as the circle, although the point that closes the curve is of a "different quality" from that of the other points, for it is the "point at infinity". The projection of a circle on a straight line shown in Figure 2.4 justifies intuitively this decision. This difference between the types of points will disappear in projective geometry, as will be seen in the following chapter. That is why the projective plane is *homogeneous*.

We call a vanishing point the **point at infinity**; that is, a "point at infinity" is where any two straight lines of the same family of parallel lines meet (there is one for each Euclidean direction). The union of all the points at infinity of a plane forms the **line at infinity**, which replaces the horizon line.

Note that the line at infinity also has the topological appearance of a circle, for the slopes of the Cartesian lines pass from $+\infty$ to $-\infty$, depending on whether the angle between the line and the positive part of the X axis approaches 90° from the right or from the left, respectively. That means that the horizon line should be sketched slightly curved, but for any practical purpose sketching it like (part of) a line is correct.

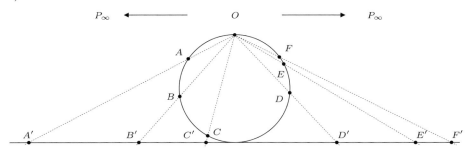

Figure 2.4: To each straight line, only one point at infinity is added.

The **affine plane**, A^2, is obtained by adding the line at infinity to the points of the Euclidean plane; considered as a set, the line at infinity remains invariant under the transformations which we shall call affinities (see Figure 2.5).

Thus, in contrast to the Euclidean plane, on the affine plane, any two lines intersect: two ordinary lines intersect at an ordinary point if their slopes are different and at a point at infinity if they have the same slope. Furthermore, an ordinary line and the line at infinity intersect at the point at infinity determined by the slope of the ordinary line.

Since we are interested in using coordinates, we must be able to assign them to the points at infinity without losing the advantage of having coordinates for the ordinary points. We know that the ordered pairs of real numbers exhaust the ordinary points; accordingly, we need another number, which can be a third coordinate, for distinguishing the ordinary points from those at infinity.

We take as a third coordinate the number 1 for the ordinary points and 0 for the points at infinity. The reason for this choice will be evident in the next chapter.

The ordinary points have as first coordinates the usual ones, so $(x, y, 1)$ are the coordinates of an ordinary point.

The points at infinity, each common to all the lines parallel to certain direction, have as first coordinates a pair of numbers λ and μ whose quotient $\frac{\mu}{\lambda}$ determines the slope of their direction. Then λ and μ cannot be both zero, but the quotient can be infinite.

That means we are dealing with a **class of pairs**: all of those which determine the same quotient. The notation $(\lambda : \mu : 0)$ for the coordinates of a point at infinity will make us remember that we must consider as equal two triples which define the same slope, such as $(2 : 3 : 0)$ and $(-4 : -6 : 0)$.

This way of taking the coordinates of a point at infinity agrees with several facts: each point at infinity defines a **class of lines**, those which are parallel to the given direction. Moreover, there are as many points at infinity as there are slopes of lines on the plane.

We can represent the affine plane using three coordinate axes, as shown in Figure 2.5: the line at infinity, characterized by the third coordinate $z = 0$, and the two ordinary axes (which are not necessarily perpendicular one to another, just transversal) characterized by $y = 0$ and $x = 0$. In these axes the third coordinate of the points must be 1.

In order to locate points on the affine plane, we need to fix the point corresponding to $(1, 1, 1)$, because then the point corresponding to 1 on the axes $y = 0$ and $x = 0$ arises from cutting them, respectively, with the lines determined by $(0 : 1 : 0)$ and $(1, 1, 1)$, and by $(1 : 0 : 0)$ and $(1, 1, 1)$. Fixing the point $(1, 1, 1)$ is equivalent to sketching the first tile in Figure 2.3. Once that is done, the points on each axis corresponding to 2, 3, and so on, are determined by the construction shown in Figure 2.3; the same can be said of the points assigned to rational numbers p/q on each axis.

By paying attention to the denominator of p/q, we can see that the last result is a consequence of the following remark: for obtaining the point of coordinates $(1/q, 0, 1)$, it suffices to draw the line through $(0, 0, 1)$ and $(1, q, 1)$, determining thus one point on the upper side of the first tile corresponding to $(1/q, 1, 1)$; the line through $(1/q, 1, 1)$ and $(0 : 1 : 0)$ intersects the lower side of the first tile at the

2.1. The line at infinity

point $(1/q, 0, 1)$. In Figure 2.5 the points of coordinates $(1/2, 3, 1)$ and $(1/3, 0, 1)$ appear.

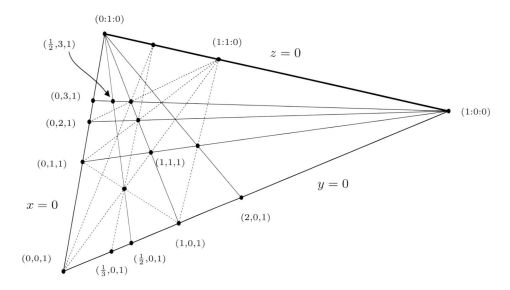

Figure 2.5: Coordinate system for the affine plane.

The points corresponding to negative numbers are determined in the same way, applying the construction for drawing tiles located to the left of (or below) the tile with vertices $(0, 0, 1)$, $(1, 0, 1)$, $(0, 1, 1)$ y $(1, 1, 1)$ (see Figure 2.6).

To obtain the remaining points corresponding to real numbers on the axes $y = 0$ and $x = 0$, it suffices to consider that a real number is the limit of a sequence of rational numbers, as in the case of the real line.

As to the line at infinity, its points also have coordinates which are determined once we have fixed the ordinary point $(1, 1, 1)$. For instance, the point at infinity of the diagonals is $(1 : 1 : 0)$ is 1; for the lines of slope $\frac{1}{2}$, the point at infinity is $(2 : 1 : 0)$; and so on.

To draw a line on the affine plane, whose equation is the same as that corresponding to the Euclidean case,

$$Ax + By + C = 0,$$

we must locate the corresponding point on the line at infinity, $(B : -A : 0)$, and take some ordinary point (for instance one of the intersections with the axes, $y = 0$ or $x = 0$). In Figure 2.6, we show the lines whose equations are $3x - y + 2 = 0$ and $-x + 2y + 4 = 0$.

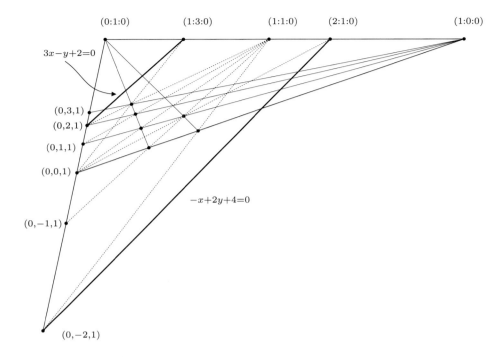

Figure 2.6: Straight lines on the affine plane.

Exercises

1. Determine the coordinates of the point at infinity of the line
$$-3x + 4y - 5 = 0.$$

2. Give the equation of the line passing through the ordinary point $(4, 1, 1)$ and the point at infinity $(3 : 2 : 0)$; also, give the equation of the line through $(0, 0, 1)$ and the same point at infinity.

3. Give a geometric algorithm for determining, on the plane of the sketch, the points corresponding to tiles having as a side the fourth part of the side of the tiles of the previous sketch. Justify your results.

4. Copy Figure 2.3 and draw the other tiles having a common side with the shadowed tile.

5. Draw the lines corresponding to the equations of Exercises 1 and 2 on an affine coordinate system given in advance on the paper of the drawing.

6. Make a construction to locate the point corresponding to the third horizontal tile on the same row of the tile sketched in the figure below.

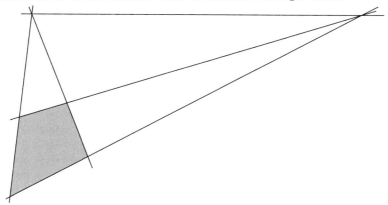

7. Sketch the representation of a rectilinear segment on the affine plane and give a geometric algorithm for obtaining the point of the sketched segment that represents the midpoint of the original segment. Justify this.

2.2 Affine transformations and their invariants

If two persons located at different positions on the same floor of mosaic tiles draw it on transparent pieces of paper, when superposing both pieces of paper, we can see that the sketching of the same marked mosaic tile has different positions on each piece of paper (the right part of Figure 2.7 shows the superposed pieces of paper). However, it is possible to go from one drawing to the other by applying a transformation, composition of a translation $T_{\bar{a}}$ and a non-singular linear transformation L, to the sides OH_1 and OV_1, as explained below.

The linear transformation L is that which takes the vectors OH_1 and OV_1 of the drawing $\mathcal{D}_1$, into the vectors $\tilde{O}\tilde{H}_1$ and $\tilde{O}\tilde{V}_1$ of the drawing $\mathcal{D}_2$. Since from no position on the floor do we see two contiguous edges as parallel, the vectors OH_1 and OV_1 are linearly independent, as well as the vectors $\tilde{O}\tilde{H}_1$ and $\tilde{O}\tilde{V}_1$. That is why L is non-singular. The translation is that which carries the point O of the drawing $\mathcal{D}_1$ into the point $\tilde{O}$ of the drawing $\mathcal{D}_2$.

Then we complete each drawing by following the same rules as above.

The previous discussion shows that this type of transformations can be useful. Let us define them formally.

An **affine transformation** is the composition of a translation $T_{\bar{a}}$ and a non-singular linear transformation L:

$$A(P) = T_{\bar{a}} \circ L(P) = L(P) + \bar{a}.$$

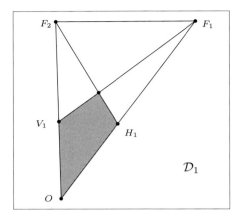 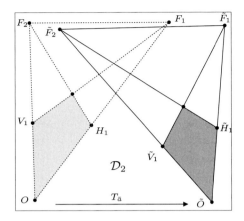

Figure 2.7: To take $\mathcal{D}_1$ into $\mathcal{D}_2$, we compose a translation with a non-singular linear transformation.

The affine transformations form a group, the **affine group**, which we denote by $A(2)$. Let us show that $A(2)$ is indeed a group.

i) The composition of two affine transformations is an affine transformation,

$$\begin{aligned} A_2 \circ A_1(P) &= (T_{\bar{a}_2} \circ L_2) \circ (T_{\bar{a}_1} \circ L_1)(P) \\ &= L_2(L_1(P) + \bar{a}_1) + \bar{a}_2 \\ &= L_2 \circ L_1(P) + (L_2(\bar{a}_1) + \bar{a}_2), \end{aligned} \qquad (2.1)$$

whose linear part is the composition of the linear parts (which implies that it is non-singular), and whose translation vector is obtained by adding $\bar{a}_2$ to the vector obtained by applying the transformation L_2 to the first translation vector, $\bar{a}_1$.

ii) Associativity holds for any transformations.

iii) The identity element can be expressed as an affine transformation by composing the translation by $\bar{0}$ with the identity element.

iv) The inverse of an affine transformation A is an affine transformation, A^{-1}, which is obtained as follows:

$$P' = L(P) + \bar{a} \;\Rightarrow\; \begin{aligned} P &= L^{-1}(P' - \bar{a}) \\ &= L^{-1}(P') + L^{-1}(-\bar{a}) = \left(T_{L^{-1}(-\bar{a})} \circ L^{-1}\right)(P'). \end{aligned}$$

Notice that the linear part of the inverse transformation is the inverse L^{-1} of the linear part L, whereas the translation on the transformation A^{-1} corresponds to the vector $L^{-1}(-\bar{a})$.

2.2. Affine transformations and their invariants

Affine transformations can be represented by 3×3-matrices such as the one below:
$$\begin{pmatrix} a & b & r \\ c & d & s \\ 0 & 0 & 1 \end{pmatrix}, \quad \text{with } ad - bc \neq 0, \tag{2.2}$$

where the linear part is given by the submatrix of entries a, b, c, d, and the translation is given by the column vector $(r, s)^t$.

Let us verify that the elements in $A(2)$ determine transformations of the affine plane A^2 having the properties one would expect from the rules of perspective.

The "type" of point, ordinary or a point at infinity, is an invariant under the affine group, for when the column vector corresponding to an ordinary point $(x, y, 1)$ is multiplied by a matrix of the form (2.2), the product is an ordinary point, that is, it has as third coordinate 1; similarly, a point at infinity is taken into a point at infinity, for the third coordinate is zero:

$$\begin{pmatrix} a & b & r \\ c & d & s \\ 0 & 0 & 1 \end{pmatrix} \begin{pmatrix} x \\ y \\ 1 \end{pmatrix} = \begin{pmatrix} ax + by + r \\ cx + dy + s \\ 1 \end{pmatrix}; \quad \begin{pmatrix} a & b & r \\ c & d & s \\ 0 & 0 & 1 \end{pmatrix} \begin{pmatrix} x \\ y \\ 0 \end{pmatrix} = \begin{pmatrix} ax + by \\ cx + dy \\ 0 \end{pmatrix}.$$

This means that parallel lines go to parallel lines, although they may have the opposite direction. Notice that if this were not true, the linear part of the affine transformation would not be injective (why?), but we have already shown that it is invertible.

The **incidence**, that is, the fact that a line goes through a point, or that a point is on a line, or that two lines intersect, is an affine invariant, namely, the transformed elements must also be incident. This property is formally trivial, but it is not so when drawing by hand.

Another affine invariant is the degree of the polynomial equation that defines a locus. A translation leaves invariant the coefficient of the term of highest degree, as the reader should prove. For the case of a non-singular linear transformation, we shall prove this fact in the following chapter for a larger group, $GL(3, \mathbb{R})$.

In order to obtain the equation of the locus $\mathcal{F}'$ into which the locus given by a polynomial equation $F(x, y) = 0$ is transformed, we proceed as always: If $P' = A(P)$, then $P = A^{-1}(P')$, and by substituting (x, y) in terms of (x', y') in the equation of the original locus, we obtain the equation of the transformed locus.

As an example, let us take a specific line and a specific conic:

$$-x + 2y - 3 = 0; \quad y^2 = 4x, \tag{2.3}$$

and let us obtain the equations of the resulting line and conic under the affine transformation

$$A = \begin{pmatrix} 1 & 1 & 2 \\ 2 & -3 & -2 \\ 0 & 0 & 1 \end{pmatrix}.$$

The inverse matrix of A is (the reader must verify it)

$$A^{-1} = \begin{pmatrix} -3 & -1 & 4 \\ -2 & -1 & 2 \\ 0 & 0 & 1 \end{pmatrix},$$

and by applying it to $(x', y', 1)$ to obtain $(x, y, 1)$ we have:

$$\begin{pmatrix} -3 & -1 & 4 \\ -2 & -1 & 2 \\ 0 & 0 & 1 \end{pmatrix} \begin{pmatrix} x' \\ y' \\ 1 \end{pmatrix} = \begin{pmatrix} -3x' - y' + 4 \\ -2x' - y' + 2 \\ 1 \end{pmatrix} = \begin{pmatrix} x \\ y \\ 1 \end{pmatrix}.$$

That is, the expressions for x and y which we must substitute in (2.4) are the following:

$$x = -3x' - y' + 4; \qquad y = -2x' - y' + 2.$$

The transformed equations of the line and the conic are, respectively,

$$\begin{aligned} -x + 2y - 3 &= -(-3x' - y' + 4) + 2(-2x' - y' + 2) - 3 \\ &= -x' - y' - 3 = 0, \\ y^2 - 4x &= (-2x' - y' + 2)^2 - 4(-3x' - y' + 4) \\ &= 4x'^2 + y'^2 + 4x'y' + 4x' - 12 = 0. \end{aligned}$$

Note that our computations showed that under an affine transformation an equation of the first degree is transformed into another equation of the first degree. Similarly, an equation of the second degree is transformed into another equation of the second degree.

One can ask whether an ellipse can be transformed into a parabola or a hyperbola under an affine transformation. We shall prove that this does not happen, for the type of conic is an affine invariant.

Since a translation does not change the type of conic (for it is a rigid transformation), it suffices to verify the statement for a non-singular linear transformation L; we do it proceeding analogously as we proved the invariance of the discriminant of a conic under orthogonal transformations.

If the conic is given by

$$Ax^2 + 2Bxy + Cy^2 + Dx + Ey + F = 0,$$

its discriminant is $B^2 - AC$, the additive inverse of the determinant of the matrix M of the quadratic part.

For the conic transformed under L, the discriminant is the additive inverse of the determinant of the matrix of the quadratic part: $(L^{-1})^t M L^{-1}$ (we have denoted by L^{-1} the matrix of the inverse transformation of L).

Since the determinant of a product is the product of the determinants, and the determinant of a matrix equals that of its transpose, the sign of the discriminant of the original conic equals that of the discriminant of the transformed conic, thus the type of conic is preserved.

2.2. Affine transformations and their invariants

This property agrees with the fact that an ellipse does not have points at infinity since the ellipse is a bounded conic. A parabola has a point at infinity ($\mathcal{P}$ in Figure 2.8), in the sense that the line tangent to the parabola at one of its points tends to become parallel to the focal axis when the point recedes indefinitely from the vertex (on either side of the focal axis). Hence, the point at infinity of the axis of the parabola is the limit of the points at infinity of the lines tangent to the parabola when the points of tangency recede from the vertex (on either side of the focal axis).

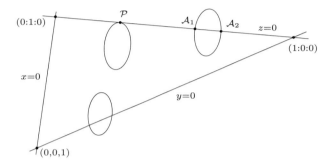

Figure 2.8: The points at infinity of each type of conic.

Similarly, a hyperbola has two points at infinity, given by the slopes of its asymptotes ($\mathcal{A}_1$ y $\mathcal{A}_2$ in Figure 2.8), for the tangent line at one of its points tends to one of the asymptotes when the point recedes indefinitely from the vertex of the branch to which it belongs. Each branch of the hyperbola is on different sides of the line at infinity, and the reader should not be surprised by the way the two branches of the hyperbola stick to each other, because a Möbius strip intervenes there (see Section 3.2).

This suggests how the concept of **asymptote** can be defined in general: an asymptote of a curve $\mathcal{C}$ in the affine plane is a line $\mathcal{L}$ (having no ordinary points of $\mathcal{C}$), such that the point at infinity of the lines tangent to $\mathcal{C}$ approaches the point at infinity of $\mathcal{L}$ as the the points of $\mathcal{C}$ approach the line $\mathcal{L}$.

That is, the concept of asymptote is a concept of affine geometry.

At this point, we hope the reader will admit a fact that may seem surprising: the sketch of each of the affine conics in the affine plane is a closed curve, for the point at infinity of the focal axis "closes" the parabola, and the points at infinity of the hyperbola "stick" one branch to the other.

We leave to the reader the task of proving that the affine group $A(2)$ has $E(2)$, the Euclidean group of rigid transformations, as subgroup. Hence $A(2)$ also contains the group of orthogonal transformations $O(2)$, the group of the rotations $SO(2)$, and the group of translations, which is isomorphic to $\mathbb{R}^2$ as an additive group. In fact there is a classical operation in group theory called the *semi-direct product* of groups, and the reader interested in this can verify that $A(2)$ is the semi-direct product of $O(2)$ and $\mathbb{R}^2$. This is in fact an easy exercise using the matrix representation of affine transformations given in (2.2).

The group of projective transformations, which we study in the next chapter, is determined by all the 3×3-matrices with non-zero determinant, so it has the affine group as a subgroup, by (2.2). Hence there are more affine invariants, in particular, all those which are invariant under the group of projective transformations.

There is another affine invariant, which is not a projective invariant, that we are interested in emphasizing: **the ratio in which a point** P **divides a rectilinear segment** AB. We know that translations preserve that ratio because they are rigid transformations, and it is easy to verify this same statement for linear transformations L.

Let us suppose that the ratio in which the point P divides rectilinear segment AB is λ, that is,

$$P - A = \lambda(B - P).$$

Then, $L(P)$ also divides the segment $L(A)L(B)$ in the same ratio, as the following computation proves:

$$L(P) - L(A) = L(P - A) = L(\lambda(B - P)) = \lambda(L(B) - L(P)).$$

This invariant is translated on the plane of perspective as a local model of the affine plane, since the corners of the tiles were determined by fixing the segment corresponding to the unit on the axes $x = 0$ and $y = 0$.

The reason for calling the plane of perspective a **local model** of the affine plane will be clear when we construct the projective plane in the next section.

Exercises

1. Find the inverse transformation of

$$T = \begin{pmatrix} 3 & 1 & -2 \\ 2 & 2 & 4 \\ 0 & 0 & 1 \end{pmatrix}.$$

2. Find the transformed point of the point at infinity $(3:2:0)$ under the transformation T of Exercise 1.

3. Apply the previous transformation to the points $A(2,4,1)$ and $B(2,8,1)$, and verify that the mid-point of the segment AB is transformed into the mid-point of the segment $T(A)T(B)$, where T is the transformation of Exercise 1.

4. Prove that the coefficient of the term of highest degree of a polynomial equation in two variables, $P(x,y)$, does not vary when a translation is applied.

2.2. Affine transformations and their invariants

5. Apply the transformation of Exercise 1 to the following conics, and verify that, in each case, they are of the same type as the original conic.
$$\frac{x^2}{4} + \frac{y^2}{16} = 1; \qquad \frac{x^2}{4} - \frac{y^2}{4} = 1.$$

6. On a long piece of transparent paper, draw a good part of the hyperbola
$$\frac{x^2}{100} - y^2 = 1,$$
and then glue the two borders of each asymptote by making one half turn in one of the short edges. You are constructing a Möbius strip.

7. Determine the subgroup of $A(2)$ corresponding to $E(2)$, and the subgroups corresponding to $O(2)$, $SO(2)$, and to the translations.

8. Define the n-dimensional Euclidean space $\mathbb{R}^n$, the dot product in $\mathbb{R}^n$, its group of isometries $E(n)$, and its subgroups: $O(n)$, $SO(n)$, and the group of translations in $\mathbb{R}^n$.

9. How many directions are there in $\mathbb{R}^3$? How many points at infinity is it necessary to add? How about $\mathbb{R}^n$?

10. Generalize the definition of the affine group to $A(n)$, ensuring that $E(n)$ is a subgroup of it.

Projective geometry

3

Projective geometry is the most extensive geometry among those we study in this book: its group of transformations admits as subgroups those we have already studied and those we shall study later on.

In addition to the Euclidean and the affine groups, there are other subgroups of the projective group playing an important role in geometry. In particular, we are interested in studying those which give rise to non-Euclidean geometries. The final part of this chapter is devoted to the so-called elliptic geometry, and Chapter 4 deals with hyperbolic geometry.

As in Chapter 2, we work mainly in dimension 2, but in some topics we shall extend our considerations to higher dimensions with no difficulty at all.

Of course, projective geometry owes its name to the fact that projections are permitted in it. Hence, in this geometry the type of conic is not an invariant, for any conic can be obtained from another by means of an adequate projection, as illustrated in Figure 3.1.

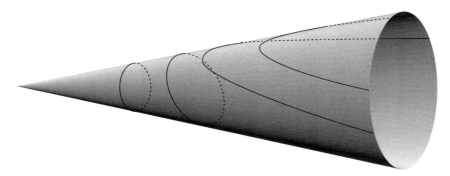

Figure 3.1: By projecting a circle on planes with different angles of inclination with respect to the axis of the cone of light another non-singular conic is obtained.

The circular border of the screen of a lamp is projected on the wall as a branch of a hyperbola, but if we interpose an inclined pasteboard, the border of the illuminated region may be a circle, an ellipse, a parabola, or a hyperbola. The singular cases — a point, a double straight line, and two intersecting straight lines — result when the plane passes through the center of projection, which is the vertex of the cone of light.

By extending the group of transformations that we consider we lose some invariants, but interesting new notions arise, such as duality, whose beauty and power are present in many branches of mathematics.

3.1 The real projective plane

To reproduce a three-dimensional object on a plane accurately, the Italian artists devised the following technique, illustrated by Albrecht Dürer (1471–1528) in [Du] (see Figure 3.2).

Figure 3.2: Renaissance technique to draw objects on a canvas.

A ring was fixed on the wall at the height of the eyes of a person, and the free extreme of a cord joined to a piece of plumb was inserted through the ring. The free extreme passed through a frame, and the cord was stretched up to touching a point of the object so that it would be visible from the ring. Finally, a horizontal cord and a vertical one were fixed to the frame in such a way that these would meet touching the cord through the frame. The meeting point of the horizontal and the vertical cords allowed marking on a cardboard (pasted on another frame) a point corresponding to the point of the object. Doing this for all points in the surface of the object, one gets a representation of it on the cardboard, a "projection" of the object on the cardboard.

The frame could be situated more or less far away from the object, and surely, it was a little bit inclined, but in any case, *each point of the tense cord represented the selected point of the object*.

3.1. The real projective plane

The way we shall define a projection is based precisely on that property of the cord.

Moreover, notice that if our eye actually were a point and we could look around in all directions, as many as diameters in a sphere, each direction would touch the objects around us in exactly one point. That is why we obtain a representation of the objects in our environment on the plane of the drawing, from which we recognize these objects.

In the chapter on Euclidean geometry we learned how to construct new spaces from others already known, by having a method to form equivalence classes of points in the given space; each such equivalence class was then considered a point of the new space.

In this spirit, to construct the projective plane we define in $\mathbb{R}^3 - \{(0,0,0)\}$ the equivalence relation suggested by the previous discussion, and which we denote by "$\sim$":

$$(x_1, x_2, x_3) \sim (x'_1, x'_2, x'_3) \tag{3.1}$$

if there is a $\lambda \in \mathbb{R} - \{0\}$ such that $x_i = \lambda x'_i$.

The geometric interpretation of this equivalence relation of the points of $\mathbb{R}^3 - \{(0,0,0)\}$ is that *every point of a straight line through the origin, except the origin itself, belongs to the same class.*

The set of all these equivalence classes has dimension 2, since we are losing one dimension when identifying each line to a point.

A **projective point** is the class of points in $\mathbb{R}^3$ which define the same direction, and the **real projective plane**, $P^2(\mathbb{R})$, consists of all the projective points, that is, the classes defined in $\mathbb{R}^3 - \{(0,0,0)\}$ by the equivalence relation (3.1).

The class defined by a non-zero triple (a, b, c) is denoted as $(a : b : c)$; the colons between the coordinates help us to remember that we are considering *classes* of triples, not the triples themselves, and we bear in mind that at least one of the coordinates is not zero.

Notice that this equivalence relation can be defined among the non-zero vectors in any vector space V over a field k (we shall deal only with $k = \mathbb{R}$ or $\mathbb{C}$): two non-zero vectors $\bar{v}$ and $\bar{w}$ are related if and only if one of them is obtained from the other by multiplication by a non-zero scalar:

$$\bar{v} \sim \bar{w} \quad \text{if and only} \quad \bar{v} = \lambda \bar{w} \quad \text{for some} \quad \lambda \in k - \{0\},$$

that is, if they belong to the same one-dimensional subspace.

The set thus obtained is called the **projectivized space** of V, denoted $P(V)$. For $V = \mathbb{R}^3$ we write

$$P^2(\mathbb{R}) = P(\mathbb{R}^3).$$

The reader should not be surprised because the numbers differ: the exponent in the first member of the equality denotes the dimension, 2, of the real projective plane, and the exponent in the second member, 3, denotes the dimension of $\mathbb{R}^3$, which is diminished in the projection by 1 because all points in the same one-dimensional subspace are equivalent, since any two of them are scalar multiples of each other.

There are two examples we are interested in analyzing: $P^1(\mathbb{R})$ and $P^1(\mathbb{C})$. Both cases are interesting in their own right; the first one because $P^1(\mathbb{R})$ is the **real projective line**, that corresponds to the "directions" of lines in the Euclidean plane, and the second one because it turns out to be "equal" to the 2-sphere (as proved below), and we shall develop the models of hyperbolic geometry in it.

A **projective point** in $P^1(\mathbb{R}) = P(\mathbb{R}^2)$ corresponds to a straight line through the origin in $\mathbb{R}^2$. Such a line meets the circle $S^1 = \{(x,y) \in \mathbb{R}^2 \mid x^2 + y^2 = 1\}$ at exactly two antipodal points (x,y) and $(-x,-y)$. Therefore the projective line $P^1(\mathbb{R})$ can be also defined as the space obtained from the circle S^1 by identifying antipodal points.

This means that in order to have a set of representative elements, it suffices to take a semi-circle and identify its extremes. The resulting figure is a closed curve, like the circle, and all the points have the same quality (since the semi-circle could be any half of S^1) in contrast with an affine straight line where the point at infinity is different from the other points.

In topology, it is said that $P^1(\mathbb{R})$ is **homeomorphic** to a circle because there exists a bijective, continuous, and invertible function between $P^1(\mathbb{R})$ and S^1, whose inverse function is likewise continuous. The projection shown in Figure 3.3 exhibits such a correspondence, a homeomorphism. In Appendix 5.5 we state precisely what is understood by a topology and by a relative topology, and we prove that this projection is in fact a homeomorphism.

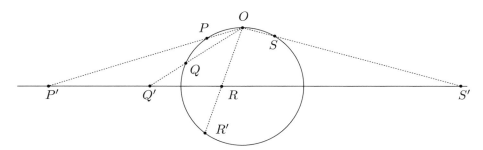

Figure 3.3: A real projective straight line is homeomorphic to a circle.

In the case of $P^1(\mathbb{C})$, which is the projectivized $P(\mathbb{C}^2)$ of $\mathbb{C}^2$, we must form classes of ordered pairs of complex numbers not both zero, $(z:w)$, and the equivalence thus obtained, $(z:w) \sim (z':w')$, takes place when there exists some complex number $\lambda \in \mathbb{C} - \{0\}$ such that $(z,w) = (\lambda z', \lambda w')$.

To determine the form of $P^1(\mathbb{C})$, we first take pairs of complex numbers (z,w) where $w \neq 0$. The division by w produces representative elements of the

3.1. The real projective plane

type $(z : 1)$, and that makes evident that there are as many classes with non-zero second coordinate as there are elements in $\mathbb{C}$.

The pairs with $w = 0$ form a single class, $(1 : 0)$, which corresponds to the only complex direction that exists in $\mathbb{C}$: any non-zero complex number z is obtained from another non-zero complex number z^* by multiplying it by the non-zero complex number z/z^*, and hence $(z : 0) = (z^* : 0)$.

The point $(1 : 0) \in P^1(\mathbb{C})$ is usually denoted by ∞, and that justifies writing $P^1(\mathbb{C}) = \mathbb{C} \cup \{\infty\}$. However, this is merely a notation, for the elements of $P^1(\mathbb{C})$ all have the same quality. Another notation is $\mathbb{C} \cup \{\infty\} = \widehat{\mathbb{C}}$, called the **extended complex plane**.

There is another way of interpreting $P^1(\mathbb{C})$. For that purpose, let us notice that $\mathbb{C}^2$ has real dimension 4 and, as in the real case, we can restrict ourselves to vectors $(z, w) \in \mathbb{C}^2$ of norm 1. That is, if $z = x + iy$ and $w = u + iv$, we shall require the real and imaginary parts to satisfy $x^2 + y^2 + u^2 + v^2 = 1$, which is the equation of S^3, the sphere of radius 1 with center at the origin:

$$S^3 = \{(x, y, u, v) \in \mathbb{R}^4 | \, x^2 + y^2 + u^2 + v^2 = 1\}.$$

The (projective) points of $P^1(\mathbb{C})$ are one-dimensional complex subspaces of $\mathbb{C}^2$ through the origin (except the origin itself),

$$\{(z, w) \in \mathbb{C}^2 - (0, 0) | \, w = kz, \text{ being } k \in \mathbb{C} \text{ fixed }\},$$

which can be interpreted as a plane through the origin of $\mathbb{R}^4$; this plane intersects S^3 in a circle whose points are representative elements of the same class (in the real case, the antipodal points of S^1 result from cutting S^1 with the straight line through the origin formed by the points in the same class). That is why we can write

$$P^1(\mathbb{C}) = P(\mathbb{C}^2) = S^3/S^1.$$

This means that $P^1(\mathbb{C})$ is obtained from S^3, identifying with a point each circle given by the intersection of S^3 with a complex line through the origin.

Regarding its form, as mentioned before, $P^1(\mathbb{C})$ is homeomorphic to the two-dimensional sphere S^2,

$$S^2 = \{(x, y, z) \in \mathbb{R}^3 | \, x^2 + y^2 + z^2 = 1\},$$

which is called the **Riemann sphere** in this context. A homeomorphism (that is, a bijective, continuous, and invertible function, whose inverse function is likewise continuous) between $P^1(\mathbb{C})$ and S^2 is obtained by the correspondence illustrated in Figure 3.4 and defined as follows. Any point P of S^2, except the north pole, N, determines with N a straight line that intersects the plane of the equator, which we identify with $\mathbb{C}$, at one single point $z \in \mathbb{C}$ that we associate with $(z : 1)$. Hence, if we agree to assign to N the point $(1 : 0)$, we obtain the so-called **stereographic projection**

$$\Pi_N : S^2 \to P^1(\mathbb{C})$$

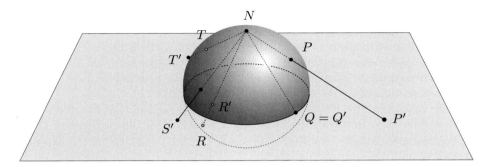

Figure 3.4: $P^1(\mathbb{C})$ is homeomorphic to S^2.

which establishes a homeomorphism between S^2 and $P^1(\mathbb{C})$ (see Appendix 5.5).

In addition, the stereographic projection has the property of being **conformal**, that is, two curves on the sphere which intersect forming a certain angle are projected into curves of the plane which intersect forming that same angle in absolute value (see Exercise 9 at the end of this section).

The constructions above generalize to $\mathbb{R}^{n+1}$ and yield a description of the n-dimensional real projective space $P^n(\mathbb{R})$ as the set of equivalence classes of the n-dimensional sphere — the unit sphere in $\mathbb{R}^{n+1}$ — where two points are equivalent if and only if they are antipodal, since there are as many directions in $\mathbb{R}^{n+1}$ as pairs of antipodal points in

$$S^n = \{\bar{x} \in \mathbb{R}^{n+1} \,|\, ||\bar{x}|| = 1\}.$$

In particular, $P^2(\mathbb{R})$ is obtained from the sphere S^2 identifying points by the **antipodal relation**:

$$\bar{x} \equiv -\bar{x}.$$

Notice that there is an obvious projection map from S^2 to $P^2(\mathbb{R})$, taking each point $x \in S^2$ into its equivalence class $[x] \in P^2(\mathbb{R})$. This map is clearly 2 to 1, and this is why we say that the sphere S^2 is a **double covering** of the projective plane $P^2(\mathbb{R})$. More generally, S^n is a double covering of $P^n(\mathbb{R})$.

Hence we can write (and it is convenient to have in mind all these interpretations of the real projective plane)

$$P^2(\mathbb{R}) = P(\mathbb{R}^3) = (\mathbb{R}^3 - \{\bar{0}\})/\sim = S^2/(\bar{x} \equiv -\bar{x}).$$

In a later section we shall deal with the topological aspect of $P^2(\mathbb{R})$, for it involves the notion of orientability, which is worthwhile discussing carefully. However, before finishing this section, we are interested in showing how the equations of the loci are established, since the coordinates are useful for that purpose.

3.1. The real projective plane

In $P^2(\mathbb{R})$, the projective lines are defined by two projective points, that is, by two linearly independent directions of $\mathbb{R}^3$; if we take one vector for each direction, the two vectors generate a plane through the origin in $\mathbb{R}^3$, that is, a subspace of dimension 2, and a projective line can be defined as follows:

A **projective line** in $P^2(\mathbb{R})$, consists of the projective points defined by coplanar directions in $\mathbb{R}^3$.

In other words, just as the points in $P^2(\mathbb{R})$ correspond to one-dimensional subspaces of $\mathbb{R}^3$, the projective lines correspond to two-dimensional subspaces of $\mathbb{R}^3$.

If we cut the sphere S^2 with a plane through the origin, the result is an equator of the sphere, that is a circle of maximal radius, and each pair of antipodal points on it must be identified to obtain representative elements of all the points forming a projective straight line. This corresponds to Figure 3.3.

The equation of a projective line is the equation of the plane through the origin of $\mathbb{R}^3$ determined by two different directions (each of them defining a projective point), $(a:b:c)$ and $(d:e:f)$.

In the three-dimensional Cartesian space, a vector normal to the plane determined by those two directions is obtained as the cross product $(a,b,c) \times (d,e,f)$, and if we denote it by (A,B,C), the equation of the plane through the origin of $\mathbb{R}^3$ is

$$Ax + By + Cz = 0.$$

The projective line defined by this equation consists of the points $(x:y:z) \in P^2(\mathbb{R})$ such that $Ax + By + Cz = 0$.

This equation is linear and homogeneous. This is precisely why if a triple (x_0, y_0, z_0) satisfies the equation, so that the corresponding point is in that plane, then also (kx_0, ky_0, kz_0) satisfies this equation for all $k \neq 0$:

$$A(kx_0) + B(ky_0) + C(kz_0) = k(Ax_0 + By_0 + Cz_0) = 0 \iff Ax_0 + By_0 + Cz_0 = 0.$$

This is a remarkable fact:

If a projective locus is given by a certain polynomial equation, then that equation must be homogeneous, for we say that a projective point satisfies the equation only if this is valid for every triple representing the point.

Notice that we are requiring the function $L(x, y, z) = Ax_0 + By_0 + Cz_0$, to take the same value (0 in this case) at equivalent points. This remark is the basis for the method used to embed $P^2(\mathbb{R})$ in $\mathbb{R}^4$ (see Section 3.3).

Now we are ready to prove the first geometric fact of projective geometry that contrasts with Euclidean geometry, for it implies that there are no parallel projective lines:

Theorem. *Every two lines of the projective plane have exactly one common point.*

Proof. Two distinct projective lines have equations

$$Ax + By + Cz = 0,$$
$$A'x + B'y + C'z = 0,$$

where (A, B, C) and (A', B', C') are non-parallel vectors in $\mathbb{R}^3$.

Each equation gives rise to a plane through the origin in $\mathbb{R}^3$, and since the planes are distinct, they intersect at a straight line through the origin with direction vector $(L, M, N) = (A, B, C) \times (A', B', C')$. That straight line through the origin corresponds to a projective point, as was stated. □

Exercises

1. Determine the point where the following projective lines intersect.
   $\mathcal{L}_1\colon 3x - y + z = 0$ and $\mathcal{L}_1\colon -x - 2y + 2z = 0$.

2. Which of the following equations define a locus in $P^2(\mathbb{R})$? Justify your answer.
   (a) $3x^2 + xy - z^2 = 0$; (b) $-x^3 - y^2 + x + z = 0$;
   (c) $x^4 + y^4 - z^4 = 0$; (d) $x^2 + y^2 + z^2 = 0$.

3. Find the representative elements of $(-2 : 6 : -4)$ having the property indicated below:
   (a) $x = 1$; (b) $z = 1$; (c) $(x, y, z) \in S^2$.

4. Prove that three projective points belong to the same projective line if and only if the determinant of the coordinates of their representative elements is zero.

5. Prove that if the coordinates $O(0 : 0 : 1)$, $H(1 : 0 : 1)$, $V(0 : 1 : 1)$, $U(1 : 1 : 1)$, are assigned to four points O, H, V, U in a **general position**, that is, such that no triple is collinear, that determines coordinates for all the other points of $P^2(\mathbb{R})$.

6. Given a point $(x_1, x_2, x_3) \in S^2$, determine the complex number $z = x + iy$ associated to it by the stereographic projection from the north pole.

7. Define $P^3(\mathbb{R})$ and explain how to characterize the projective lines and the projective planes in this projective space.

8. Define $P^n(\mathbb{C}) = P(\mathbb{C}^{n+1})$ and consider

   $$||(z_0, z_1, \ldots, z_n)|| = ||(x_0, y_0, x_1, y_1, \ldots, x_n, y_n)||,$$

   where $z_k = x_k + iy_k$ and $||(x_0, x_1, \ldots, x_n, y_n)||$ is the norm defined by the scalar product in $\mathbb{R}^{n+1}$. Prove that $P^n(\mathbb{C}) = S^{2n+1}/S^1$. (The natural projection $\Pi : S^{2n+1} \to P^n(\mathbb{C})$ is known as **the Hopf fibration**; see [DoC].)

9. Prove that the stereographic projection is **conformal**, that is, two curves on the sphere that intersect at a point P forming an angle α are mapped under the stere graphic projection into curves forming that same angle (in absolute value). (Suggestion: Consider the figure below, where τ is

3.2. The Duality Principle

the plane tangent to the sphere at the point P, $\bar{v}$ and $\bar{w}$ are the vectors tangent to the curves on the sphere; H and H' belong to the orthogonal projection of the straight line NP into the plane τ: H is the intersection of that projection and the plane tangent to the sphere at N, and H' belongs to the plane of the equator, which is parallel to the former.)

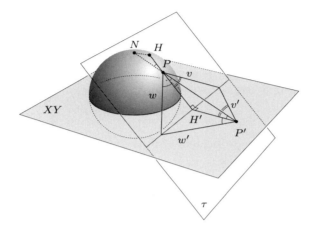

3.2 The Duality Principle

As a consequence of the theorem in the previous section, there are no exceptional cases in the projective plane for the intersection of projective lines, and that is why to the valid statement

Two points determine a line,

there corresponds another statement which is also valid:

Two lines determine a point.

This fact is the key to one of the most beautiful properties of the real projective plane, **duality**, the bijective correspondence between points and lines of the real projective plane that respects **incidence**: a point and a line are incident if the point belongs to the line, or the line passes through the point.

The **Duality Principle** of projective geometry states:

*If a valid statement is posed only in terms of points, lines and the incidence relations, then the **dual proposition**, obtained from the original statement by interchanging the terms point and line, is also true.*

For instance, if we think of a **triangle** as being the figure determined by three non-collinear points (the vertices of the triangle), then the lines determined by (that are incident with) two of those points are the three sides of the triangle.

The dual notion is that of a **trilateral**: the figure determined by three non-concurrent lines, the sides of the trilateral. The points where two sides intersect are the vertices of the trilateral.

Thus, a trilateral also has three vertices and three sides, and that justifies saying that a triangle is self-dual. However, that does not happen with a quadrangle (see Exercise 1).

The scalar product in $\mathbb{R}^3$ yields to an easy expression of the incidence of a point and a line. If the equation of the projective line is $Ax + By + Cz = 0$, determined by the (equivalence class of the) non-zero triple (A, B, C), then using the scalar product in $\mathbb{R}^3$ we can write

$$(A, B, C) \cdot (x, y, z) = 0, \tag{3.2}$$

and the point $(x_0 : y_0 : z_0)$ is **incident** with the line defined by $(A : B : C)$ if and only if

$$(A, B, C) \cdot (x_0, y_0, z_0) = 0.$$

However, this equation can also be interpreted as the incidence of the projective line defined by $(A : B : C)$ and the point $(x_0 : y_0 : z_0)$.

The geometric duality justifies the name **dual space** V^* of a vector space V given to the set of linear transformations $T : V \to k$ where k is the field of scalars (see [B-ML] or [Ri]). In fact, when $V = \mathbb{R}^3$, the triple (A, B, C) can be seen as the linear transformation $(x, y, z) \mapsto (Ax + By + Cz)$ from $\mathbb{R}^3$ into $\mathbb{R}$, and the equation (3.2) characterizes the **kernel** of that transformation, that is, the locus of the triples (x, y, z) which under the transformation defined by

$$T(x, y, z) = (A, B, C) \cdot (x, y, z)$$

are mapped into $0 \in \mathbb{R}$.

That locus is a plane through the origin of $\mathbb{R}^3$, which defines a projective line, and the correspondence is established by the class $(A : B : C)$, since (A, B, C) and $(\lambda A, \lambda B, \lambda C)$ have the same kernel.

Hence, in $P^2(\mathbb{R})$ there is a bijection between points and lines; any statement of incidence among points and lines can be dualized, and if the original statement is valid, its dual is also valid.

For example, we have already asked the reader to prove the following statement (see Exercise 4):

Statement. *Three points P, Q and R are incident with the same line if and only if the determinant of their coordinates becomes zero.*

By exchanging the terms "point(s)" and "line(s)," we obtain the dual statement:

Statement. *Three lines $\mathcal{P}$, $\mathcal{Q}$ and $\mathcal{R}$ are incident with the same point if and only if the determinant of their coordinates becomes zero.*

3.2. The Duality Principle

In both cases, we are saying that there is a linear dependency among the three objects, points in one case, lines in the other; in addition, if two of them satisfy (3.2), a linear combination of them also does.

A line can be seen as a **range of points**: all the points that are incident with the line, and a point can be seen as a **pencil of lines**: all the lines that are incident with the point. (See Figure 3.5.)

It is worthwhile mentioning that, in more colloquial language, the lines that are incident with the same point are called concurrent, and the points that are incident with the same line are called collinear.

In both cases, the pencil and the range are determined by two of its elements, and the remaining elements are combinations of those two; we express this as a remark.

Remark. If in a pencil of lines or in a range of points we choose two elements, A and B, then any other element P is a linear combination of A and B, $P = \lambda A + \mu B$, and P can be represented by $(\lambda : \mu)$.

Figure 3.5: A line is a range of points, and a point is a pencil of lines.

The most remarkable example of applying the principle of duality is the so-called Theorem of Desargues and its dual, which in this case coincides with the converse theorem.

Theorem of Desargues. *If the triangles A, B, C and A', B', C' are such that the lines defined by corresponding vertices are not collinear, then the intersections of corresponding sides are collinear.*

In Figure 3.6, the thick lines AA', BB' and CC' are incident with a point O, and the points $R = AB \cap A'B'$, $S = BC \cap B'C'$ and $T = CA \cap C'A'$ seem to be collinear; we shall prove that this is actually the case by using a technique of Desargues, of projection and section, with which he obtained a large number of results.

Before giving the proof, we shall write the dual proposition according to the rule of exchanging the terms point and line (the readers must convince themselves that we have proceeded correctly).

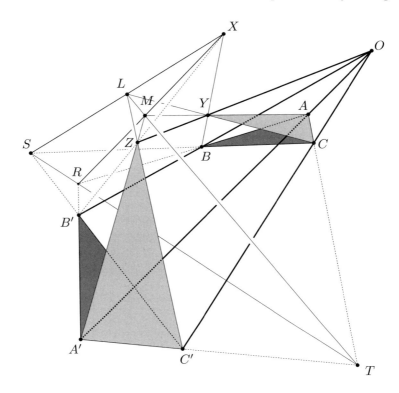

Figure 3.6: Two triangles in perspective from a point are also in perspective from a line.

Dual proposition of the Theorem of Desargues. *If the trilateral figures abc and $a'b'c'$ are such that the points defined by corresponding sides, aa', bb' and cc', are incident with a line $\mathcal{L}$, then the lines defined by corresponding vertices, $a \cap b$ and $a' \cap b'$; $b \cap c$ and $b' \cap c'$; $c \cap a$ and $c' \cap a'$, are incident with a point O.*

The two statements can be summarized thus:

Two triangles are in perspective from a point if and only if they are in perspective from a line.

Proof of the Theorem. The points O, A, B, C, A', B', C', R, S and T belong to the same plane, whereas X is a point outside that plane; in the line XB we choose a point Y, and we call Z the intersection of XB' with the line OY (the lines intersect because the points X, B, O, Y and B' are coplanar).

The lines OA, OC and OY are the edges of a triangular pyramid, and the triangles AYC and $A'ZC'$ belong to two plane sections of the pyramid. The planes of those sections intersect at a line to which $T = AC \cap A'C'$, $CY \cap C'Z = L$, and $YA \cap ZA' = M$, belong, and when we project the triangles ACY and $A'C'Z$ into the original plane, we retrieve the triangles ABC and $A'B'C'$.

The projection of the collinear points T, L and M in the original plane gives rise to the points, necessarily collinear, T, S and R. □

The relation between the projective plane and the affine plane will be clear below, but for now we want to note that each projective point $(x : y : z)$ admits a representative element where at least one of the coordinates is 1, for

- if $x \neq 0$, $(x : y : z) = (1 : y/x : z/x)$;
- if $y \neq 0$, $(x : y : z) = (x/y : 1 : z/y)$;
- if $z \neq 0$, $(x : y : z) = (x/z : y/z : 1)$,

and in the third form we already recognize the coordinates of the ordinary points of the affine plane.

Exercises

1. Dualize the concept of quadrilateral, given at the beginning of Chapter 2.

2. Justify the following statement:
   In $P^3(\mathbb{R})$, a tetrahedral is self-dual.

3. Prove the Theorem of Desargues using Exercise 5 of Section 3.1.

4. Generalize to $P^n(\mathbb{R})$ the notion of duality. That is, establish a bijection between the set of points $(x_0 : x_1 : \ldots : x_n)$ and that of the projective hyperplanes, such that it respects the incidence, understanding by projective hyperplane the subset defined by an $(n + 1)$-tuple, $(A_0 : A_1 : \cdots : A_n)$ by means of the equation
$$(A_0 : A_1 : \cdots : A_n) \cdot (x_0 : x_1 : \cdots : x_n) = 0.$$

5. Use the scalar product to prove that in $\mathbb{R}^{n+1}$ there is a duality between the set $G_{k,n}$ of the k-subspaces and the set $G_{(n-k),n}$ of the $(n-k)$-subspaces for every $k \in 1, 2, \ldots, n$. Those sets are called **Grassmannian manifolds** (see [Mat]), and with them a duality is established in $P^n(\mathbb{R})$ between the projective $(k-1)$-planes and the projective $(n-k-1)$-planes.

3.3 The shape of $P^2(\mathbb{R})$

After defining the real projective plane, we write

$$P^2(\mathbb{R}) = P(\mathbb{R}^3) = (\mathbb{R}^3 - \{\bar{0}\})/\sim \, = S^2/(\bar{x} \equiv -\bar{x}),$$

and the last expression indicates that when we restrict ourselves to unit vectors in $\mathbb{R}^3$, that is, to the points of the sphere, we identify antipodal points to get the projective plane. Hence, we can retain the points of a hemisphere whenever we identify antipodal points on the boundary.

Figure 3.7 suggests that the gluing cannot be done in $\mathbb{R}^3$ if we do not allow the surface that we construct to have a self-intersection. But points of self-intersection should not exist, because although each point of the arc AB is identified with a point of the arc $A'B'$, and each point of the arc BA' is identified with a point of the arc $B'A$, there are no points of contiguous arcs which should be identified, since they are not antipodal. However, those points necessarily arise if we want to construct $P^2(\mathbb{R})$ in $\mathbb{R}^3$.

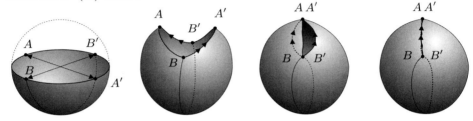

Figure 3.7: $P^2(\mathbb{R})$ cannot be constructed in $\mathbb{R}^3$ without self-intersections.

Instead, $P^2(\mathbb{R})$ fits into $\mathbb{R}^4$ without self-intersections. To prove it, we define $F : P^2(\mathbb{R}) \to \mathbb{R}^4$ by

$$(x : y : z) \mapsto (x^2 - y^2, xy, xz, yz).$$

Any two antipodal points of the sphere are mapped into the same point of $\mathbb{R}^4$, and that allows us to define a function in $P^2(\mathbb{R})$, $F : P^2(\mathbb{R}) \to \mathbb{R}^4$, given by

$$(x : y : z) \mapsto (x^2 - y^2, xy, xz, yz),$$

whose injectivity (which the reader must verify) allows $P^2(\mathbb{R})$ to fit in $\mathbb{R}^4$ without self-intersections.

The projective plane is a surface with only one side. To understand what this means, we take again the sphere and mark on it a symmetric belt around a great circle (see Figure 3.8).

It is clear that we can eliminate one of the two caps, for the points of a cap have their antipodal points in the other cap, and it is also true that we can eliminate half of the belt, for example that which is behind, since the corresponding antipodal points are in the front half of the belt.

The left and right edges of the front half of the belt are formed by antipodal points and, therefore, they must be identified in the way indicated by the arrows; that can be done only after twisting one short edge 180°, originating a **Möbius strip**, named after its creator, F. A. Möbius (1790–1868).

To finish constructing the projective plane, the border of the Möbius strip must be pasted to the border of the disk, but again we shall have problems to do the gluing in our three-dimensional space.

3.3. The shape of $P^2(\mathbb{R})$

Figure 3.8: $P^2(\mathbb{R})$ contains a Möbius strip.

The Dutch engraver Maurits Cornelis Escher (1898–1972) [Es] made several engravings that illustrate the essential property of a Möbius strip: one can walk on the Möbius strip in such a way that when getting back to the point where the route was initiated, our head will be toward the opposite direction, which is represented by the pin of Figure 3.9.

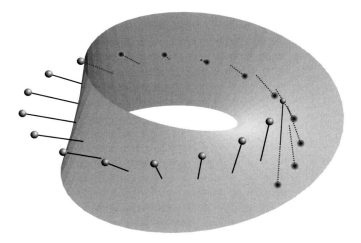

Figure 3.9: A cylinder has two sides, a Möbius strip has only one side.

To establish what we understand by orientability of a surface, we begin by defining what it means to triangulate it.

The sphere of Figure 3.10 is divided into eight "triangles," so that the following conditions hold:

i) the union of all the triangles is the sphere, and

ii) the intersection of two non-disjoint triangles can only be a common vertex or a complete common side.

The triangulation of a surface will be necessary to define the Euler characteristic of a surface; remember that in Chapter 1 we learned how to compute it for the Platonic solids and also for the torus and the double torus.

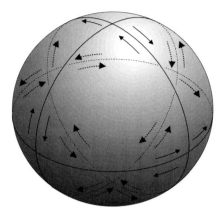

Figure 3.10: Triangulation of a sphere using oriented triangles.

We say that a **surface is orientable** if after triangulating it and drawing in one of the triangles a curved arrow indicating the direction of traversing its boundary, then when translating the curved arrow to the other triangles a common side of two triangles is always traversed in opposite directions, as illustrated in the sphere of Figure 3.10.

We invite the reader to try do that on the Möbius strip; in Figure 3.11 a strip is included before its short edges are glued, which shows why that is impossible.

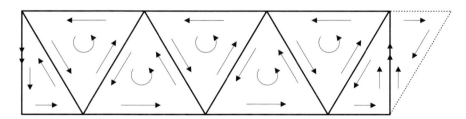

Figure 3.11: The Möbius strip is not orientable.

The arrow of the first triangle has been translated and marked in each triangle, but after identifying the edges to form the Möbius strip, we obtain one side that is traversed in the same direction for the two triangles to which that side belongs.

If in Figure 3.9 instead of the pin we had taken a screw with a right-hand spiral groove to complete a right-hand coordinate system at each point of the strip, we would have an incompatibility when crossing the side where we made the

3.3. The shape of $P^2(\mathbb{R})$

gluing; that is why we say that the Möbius strip is not orientable.

This theme deserves several additional comments; for instance, topologists construct the real projective plane from a square whose edges must be identified as indicated by the arrows in Figure 3.12 (since a square and a hemisphere are homeomorphic, in Figure 3.12 we actually have a repetition of the first part of Figure 3.7). It is clear that by gluing the edges with only one arrow, we already have a Möbius strip, that is why the gluing of the double arrows is impossible without self-intersections of the object.

The notion of orientability can be defined for objects with dimensions different from 2, and it is possible to prove that $P^n(\mathbb{R})$ is orientable if and only if n is even (see [Hr]).

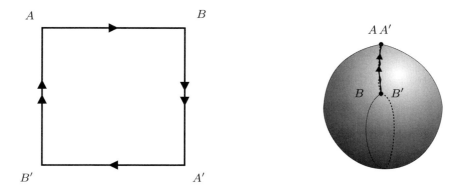

Figure 3.12: Construction of the real projective plane from a square.

Finally, it is worthwhile proving that $P^3(\mathbb{R})$ is homeomorphic to $SO(3)$, the group of rotations of $\mathbb{R}^3$, regarded as a space.

For this we notice that $P^3(\mathbb{R})$ results from identifying antipodal points of the sphere S^3, consisting of points in $\mathbb{R}^4$. This determines a topology on $P^3(\mathbb{R})$. Furthermore, to construct $P^3(\mathbb{R})$ it is obviously enough to consider the upper hemisphere of S^3, which is a 3-disc D^3, and identify the points of its boundary, which is an equator of S^3.

It is easy to see that a hemisphere of S^2 is homeomorphic to a disc D^2, the set of the points of $\mathbb{R}^2$ with norm strictly less than 1; for this it suffices to consider the orthogonal projection from the north hemisphere on the XY plane (see Figure 3.13). Analogously, a "hemisphere" of S^3 is homeomorphic to a 3-disc D^3, the points of $\mathbb{R}^3$ with norm less than 1 (in this case we project orthogonally the north "hemisphere" of S^3 upon the XYZ space), and it is immediately verified that the boundary of D^3 is S^2.

That means that we can think of $P^3(\mathbb{R})$ as the space obtained from the three-dimensional ball D^3 when we identify the antipodal points in its boundary (cf. Exercise 8).

On the other hand, we know that a rotation about the origin of $\mathbb{R}^3$ (an element of $SO(3)$ leaves a line fixed (pointwise) — the axis of the rotation — and leaves the planes orthogonal to the axis invariant (in this case as a set); all the points of those planes turn around by the same angle, that of the rotation.

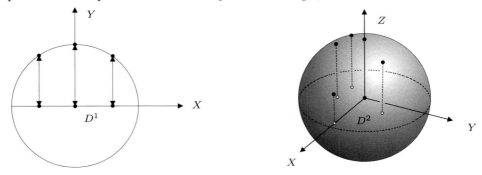

Figure 3.13: A hemisphere of S^n is homeomorphic to D^n.

There are as many possible axes of rotation as diameters of a sphere, and as many possible angles of rotation as points in the interval $[-\pi, \pi]$, except that once the axis has been chosen, rotating by π has the same effect as rotating by $-\pi$.

Thus, in the sphere of $\mathbb{R}^3$ with the center at the origin and radius π, the diameters correspond to the possible axes of rotation, and the points that lie inside that sphere, which necessarily belong to some diameter, determine an angle of rotation defined by the point of the interval $(-\pi, \pi)$ which corresponds to the diameter.

All the previous discussion means that we can identify each point that lies inside the sphere of radius π, i.e., a point of D^3_π, with a rotation whose axis contains the diameter defined by the point, and whose angle is defined by the oriented distance $\theta \in (-\pi, \pi)$ from the point to the center of the sphere. Instead, the points of the boundary, which is the sphere with center at the origin and of radius π, must be identified if they are antipodal, as occurs in the construction of $P^3(\mathbb{R})$.

Therefore, $SO(3)$ is the space resulting from D^3_π when we identify the antipodal points in its boundary, and since D^3_π and D^3 are homeomorphic, so are $SO(3)$ and $P^3(\mathbb{R})$.

Exercises

1. Construct three Möbius strips using three strips of paper. Color the first one with any color, and cut the other two this way: one along the central line, and the other along a curve equidistant from the long edge and the central line. Explain what happens in each case.

2. Draw an interval, divide it into several subintervals and, after assigning a route direction in the first one (indicate it by an arrow), copy it for the others by translating the arrow. Verify that it is possible to complete the vector associated with each arrow with another vector in the Euclidean plane, forming a right-hand base for the plane at the point, so that this choice is compatible after identifying the borders of the original segment.

3. Verify that S^2 is orientable following any of these two procedures:
   a) "Triangulate" the sphere, and verify that it is possible to orient the triangles so that the common sides are traversed in opposite directions.
   b) Compute the gradient of the function $F(x,y,z) = x^2 + y^2 + z^2$ (see Appendix 5.1) and verify that at any point $(x_0, y_0, z_0) \in S^2$, $\nabla F(x_0, y_0, z_0) \neq (0,0,0)$. The fact that $\nabla F(x,y,z)$ varies continuously with the point $(x,y,z) \in S^2$ implies that compatible orientations for curves around close points can be defined (see [DoC] for another definition of orientability).

4. Mark the horizontal sides of a square with arrows pointing to the right, and the vertical sides with arrows pointing one downward and the other upward. First glue the horizontal sides making the arrows coincide (What is obtained?), and explain why in $\mathbb{R}^3$ it is not possible to do the missing gluing by making the vertical arrows coincide. The object obtained from that identification on the sides of a square is called a **Klein bottle**, which is a non-orientable surface.

5. Rotate the circle on the $YZ \subset \mathbb{R}^3$ with center at $(0,2,0)$ and radius 1 about the axis Z. Prove that the torus of revolution T thus obtained is invariant under the antipodal mapping, and that by identifying the antipodal points of T, a Klein bottle is also obtained.

6. Prove the injectivity of the function $F : P^2(\mathbb{R}) \to \mathbb{R}^4$ given by $(x:y:z) \mapsto (x^2 - y^2, xy, xz, yz)$.

7. Give a continuous, injective, and invertible function, whose inverse function is likewise continuous, between the regions of the plane bounded by a square and by a circle.

8. Define a topology for $SO(3)$ and another for $P^3(\mathbb{R})$ (see Appendix 5.5).

9. The edge of a cap of Figure 3.8 can be deformed into a point of the cap without leaving the cap. Is it possible to deform the central line of the strip (which is a projective line) into the edge of the strip? In other words, is it possible to deform a projective line into a point without leaving $P^2(\mathbb{R})$?

10. Define $P^n(\mathbb{R})$ and assign a topology to it.

3.4 Coordinate charts for $P^2(\mathbb{R})$ (and for $P^1(\mathbb{C})$)

At the end of Section 3.2 we saw that any point of the projective plane admits representative elements of at least one of the following forms:

- if $x \neq 0$, then $(x:y:z) = (1 : y/x : z/x)$;
- if $y \neq 0$, then $(x:y:z) = (x/y : 1 : z/y)$;

- if $z \neq 0$, then $(x : y : z) = (x/z : y/z : 1)$;

that is why in differential geometry it is said that the mappings from $\mathbb{R}^2$ into $P^2(\mathbb{R})$

$$\begin{aligned} \bar{x}_1(u,v) &= (1 : u : v), \\ \bar{x}_2(u,v) &= (u : 1 : v), \\ \bar{x}_3(u,v) &= (u : v : 1), \end{aligned} \qquad (3.3)$$

form an **atlas** for $P^2(\mathbb{R})$ (see [DoC]).

Each of the mappings $\bar{x} : \mathbb{R}^2 \to P^2(\mathbb{R})$ is a **parametrization** of a region $U \subset P^2(\mathbb{R})$, and since the inverse mapping $\bar{x}^{-1} : U \to \mathbb{R}^2$, allows assigning coordinates (u,v) to a point P of the projective plane, it is called a **coordinate chart**.

If a point admits more than one of those representations, as occurs for $(2 : 3 : 4)$, the **change of coordinates** is differentiable because

$$(2 : 3 : 4) = \bar{x}_1(u_1, v_1) = \bar{x}_2(u_2, v_2) = \bar{x}_3(u_3, v_3)$$

implies that

$$(2 : 3 : 4) = (1 : u_1 : v_1) = (u_2 : 1 : v_2) = (u_3 : v_3 : 1). \qquad (3.4)$$

From the first equality of (3.4) it turns out that $(u_2, 1, v_2) = u_2(1, u_1, v_1)$, and therefore $u_2 u_1 = 1$ and $u_2 v_1 = v_2$. The change of coordinates is given by

$$u_2 = u_1^{-1}, \quad v_2 = v_1 u_1^{-1},$$

which are differentiable functions of u_1 and v_1 at $(u_1, v_1) = (3/2, 4/2)$.

The second equality of (3.4) implies that $(u_3, v_3, 1) = v_3(u_2, 1, v_2)$; accordingly, $v_3 v_2 = 1$ and $v_3 u_2 = u_3$. The change of coordinates is given by

$$u_3 = u_2 v_2^{-1}, \quad v_3 = v_2^{-1},$$

which are differentiable functions of u_2 and v_2 at $(u_2, v_2) = (2/3, 4/3)$.

The reader will have no difficulty in verifying that u_3 and v_3 are also differentiable functions of u_1 and v_1 at $(u_1, v_1) = (3/2, 4/2)$.

Also since the Jacobians $\frac{\partial(u_i, v_i)}{\partial(u_j, v_j)}$ are different from zero for all the points in the intersection $\bar{x}_i(\mathbb{R}^2) \cap \bar{x}_j(\mathbb{R}^2)$, the Theorem of the Inverse Function states that (u_j, v_j) are also differentiable functions of (u_i, v_i).

The fact that any point of the real projective plane belongs to at least one parametrized neighborhood, $\bar{x}_1(\mathbb{R}^2)$, $\bar{x}_2(\mathbb{R}^2)$, or $\bar{x}_3(\mathbb{R}^2)$, and that the changes of coordinates $(u_i(u_j, v_j), v_i(u_j, v_j))$ are **diffeomorphisms**, that is, differentiable functions with differentiable inverse function, what we have just proven is expressed by saying that $P^2(\mathbb{R})$ is a **differentiable manifold** of dimension 2.

Notice that each parametrization covers the projective plane, except for a projective line characterized by $x = 0$ in the case of $\bar{x}_1$, $y = 0$ in the case of

3.4. Coordinate charts for $P^2(\mathbb{R})$ (and for $P^1(\mathbb{C})$)

$\bar{x}_2$, and $z = 0$ in the case of $\bar{x}_1$; two of the parametrizations are not enough because the point of intersection of the two missing lines, one in each parametrized neighborhood, is missing in both.

The last parametrization given in (3.3) is the one we used for the ordinary points of the affine plane; if we had chosen any of the other two, the line at infinity would have as an equation $x = 0$ in the first case, or $y = 0$ in the second case.

Actually, it makes sense to say that the affine plane results from choosing a line of the projective plane, which must remain invariant under the allowed transformations of affine geometry.

This justifies the technique of **homogenizing a polynomial equation** with real coefficients in two variables, $F(x,y) = 0$. Let us consider as an example the Cartesian equation of a conic:

$$Ax^2 + 2Bxy + Cy^2 + 2Dx + 2Ey + F = 0. \tag{3.5}$$

If $z \neq 0$, the substitution $x \mapsto x/z$, $y \mapsto y/z$ in (3.5) gives rise to the equation

$$A(x/z)^2 + 2B(x/z)(y/z) + C(y/z)^2 + 2D(x/z) + 2E(y/z) + F = 0, \tag{3.6}$$

which is transformed into a homogeneous polynomial equation in the three variables x, y and z when we multiply by z^2:

$$Ax^2 + 2Bxy + Cy^2 + 2Dxz + 2Eyz + Fz^2 = 0. \tag{3.7}$$

This is the projective equation of a conic, and it will be easy to prove with (3.7), once we introduce the group of transformations whose invariants we are interested in, the statement made in the introduction to this chapter: that all non-singular conics are projectively equivalent.

For now notice that any projective locus given by a homogeneous polynomial equation can be studied using three affine equations, because there are three standard forms for "unhomogenizing".

In our example, if $z \neq 0$, we can divide by z^2 and retrieve the original equation (3.5), but we can also divide by y^2 for the points $(x:y:z)$ with $y \neq 0$, and obtain the equation

$$Ax^2 + 2Bx + C + 2Dxz + 2Ez + Fz^2 = 0 \tag{3.8}$$

where we write x instead of x/y and z instead of z/y.

Of course, we can proceed analogously for x: divide by x^2, which makes sense for all the points $(x:y:z) \in P^2(\mathbb{R})$ such that $x \neq 0$; in this case we obtain a second degree equation in y and z which we shall not write.

As we already stated, homogenization is a technique that we apply when we want to study projective properties of a geometric object.

In the case of $P^1(\mathbb{C})$ we can also talk about coordinate charts, but now the coordinates are complex.

We know that any point $(z:w) \in P^1(\mathbb{C})$ admits a representative element of at least one of the two following forms:

- if $z \neq 0$, $(z:w) = (1:w/z)$;
- if $w \neq 0$, $(z:w) = (z/w:1)$.

Thus, in this case two parametrizations that map $\mathbb{C}$ into $P^1(\mathbb{C})$ are enough to have all the points of $P^1(\mathbb{C})$ parametrized:

$$\begin{aligned} \bar{x}_1(z_1) &= (1:z_1), \ z_1 \in \mathbb{C} \\ \bar{x}_2(z_2) &= (z_2:1), \ z_2 \in \mathbb{C}. \end{aligned} \qquad (3.9)$$

At each parametrized neighborhood only one point is missing: $(0:w)$ for the first one, and $(z:0)$ for the second, but this point is included in the other parametrized neighborhood. For the points $(z:w)$ included in both charts, the change of coordinates is given by $z_1 = z_2^{-1}$.

Also in this case the change of coordinates is differentiable, for the derivative of a function $f: \mathbb{C} \to \mathbb{C}$ is defined exactly the same way as in the case of $\mathbb{R}$: it is the limit (if it exists) of the quotient of the increment of the value of the function and the increment of the variable, when the latter approaches zero (see [A] or [Ma]).

Hence the change of coordinates $z_1 = z_2^{-1}$ is differentiable at every point of its domain (defined by $z_1, z_2 \neq 0$), as well as its inverse function; in analysis of a complex variable, a differentiable function is called **holomorphic or analytic**, and when the inverse function is also differentiable, the original function is called a biholomorphism.

Because of this property of the change of coordinates, it is said that $P^1(\mathbb{C})$ is a **Riemann surface**. The reader interested in this theme, into which we shall go further in Chapter 4, can see [F] or [Spr].

Exercises

1. Prove that the sphere $S^2 = \{(x,y,z) \in \mathbb{R}^3 \mid x^2 + y^2 + z^2 = 1\}$ is a differentiable manifold of dimension 2, by using two parameterizations: the stereographic projection from the north pole, and the stereographic projection from the south pole. Determine the intersection of the regions of the sphere parametrized by each, and explicitly give the changes of coordinates to prove that they are diffeomorphisms.

2. Define two stereographic projections for the sphere of dimension n, $S^n = \{(x_1, \ldots, x_{n+1}) \in \mathbb{R}^{n+1} \mid x_1^2 + \cdots + x_{n+1}^2 = 1\}$, to prove that it is an n-dimensional manifold.

3. Prove that the Klein bottle is a differentiable manifold of dimension 2.

4. Prove that $P^n(\mathbb{R})$ is a differentiable manifold of dimension n. What can you say in this respect about $P^n(\mathbb{C})$?

3.5 The projective group

We have promised two things with respect to the group of transformations whose invariants we shall study in this chapter: that it will contain the affine group as a subgroup, and that, honoring its name, it will include the projections of a projective line into another from a point exterior to both lines.

As to the second point, the projections of a line into another from a point exterior to both are called **perspectivities**. Since the composition of two perspectivities is not necessarily a perspectivity (see Figure 3.16), but it must be an element of the projective group, any element of the projective group will be called a **projectivity**.

As to extending the group $A(2)$ of the affinities, which are given by invertible matrices of 3×3 of a certain type, it is natural to consider the **linear group** of order 3, $GL(3, \mathbb{R})$, consisting of all 3×3 non-singular matrices. Let us see if it is the adequate group for our purposes.

First of all, projectivities must map projective points into projective points and projective lines into projective lines. This is fine because any $T \in GL(3, \mathbb{R})$ is a non-singular linear transformation of $\mathbb{R}^3$ and therefore carries one-dimensional and two-dimensional subspaces of $\mathbb{R}^3$ (lines through the origin and planes through the origin) into subspaces of the same dimension.

However, we need more: the resulting subspace must be independent of the particular vector (or vectors) that defines the original subspace. That this is the case follows from the linearity of the function:

$$\begin{pmatrix} a & b & c \\ d & e & f \\ g & h & i \end{pmatrix} \begin{pmatrix} \lambda x \\ \lambda y \\ \lambda z \end{pmatrix} = \begin{pmatrix} \lambda(ax + by + cz) \\ \lambda(dx + ey + fz) \\ \lambda(gx + hy + iz) \end{pmatrix}.$$

Thus, any element of $GL(3, \mathbb{R})$ maps projective points into projective points and also projective lines are transformed into projective lines, because of duality (think of the vector in $\mathbb{R}^3$ normal to the two-dimensional subspace corresponding to the projective line that must be transformed).

Finally, notice that any non-zero multiple of a given matrix $M \in GL(3, \mathbb{R})$ defines the same projectivity, for the resulting triples differ only by that multiple; therefore, the obtained projective point is the same and, by duality, the same can be said of a projective line.

That is why the **projective group** will be $PGL(3, \mathbb{R})$, the set of the classes of matrices in $GL(3, \mathbb{R})$ defined by the equivalence relation

$$\begin{pmatrix} a & b & c \\ d & e & f \\ g & h & i \end{pmatrix} \sim \begin{pmatrix} a' & b' & c' \\ d' & e' & f' \\ g' & h' & i' \end{pmatrix}$$

if there exists $\lambda \in \mathbb{R} - \{0\}$ such that
$$\begin{pmatrix} a' & b' & c' \\ d' & e' & f' \\ g' & h' & i' \end{pmatrix} = \begin{pmatrix} \lambda a & \lambda b & \lambda c \\ \lambda d & \lambda e & \lambda f \\ \lambda g & \lambda h & \lambda i \end{pmatrix}.$$

It is interesting to see how the image of a point and the image of a line under the same projectivity are related.

Let us take as an equation of the line $Ax + By + Cz = 0$, and let us rewrite it as a product of matrices

$$\begin{pmatrix} A & B & C \end{pmatrix} \begin{pmatrix} x \\ y \\ z \end{pmatrix} = 0. \qquad (3.10)$$

Since the matrix M that defines a projectivity is invertible, the relation $P' = MP$, implies that $P = M^{-1}P$, and by substituting

$$\begin{pmatrix} x \\ y \\ z \end{pmatrix} = \begin{pmatrix} a & b & c \\ d & e & f \\ g & h & i \end{pmatrix}^{-1} \begin{pmatrix} x' \\ y' \\ z' \end{pmatrix},$$

in (3.10), we obtain the equation of the line transformed under the projectivity corresponding to M:

$$\begin{pmatrix} A & B & C \end{pmatrix} \begin{pmatrix} a & b & c \\ d & e & f \\ g & h & i \end{pmatrix}^{-1} \begin{pmatrix} x' \\ y' \\ z' \end{pmatrix} = 0. \qquad (3.11)$$

and $(A'\ B'\ C') \neq (0\ 0\ 0)$ because M is non-singular and $(A\ B\ C) \neq (0\ 0\ 0)$.

From this equation it is clear that the triple defining the transformed line is given as

$$\begin{pmatrix} A' & B' & C' \end{pmatrix} = \begin{pmatrix} A & B & C \end{pmatrix} \begin{pmatrix} a & b & c \\ d & e & f \\ g & h & i \end{pmatrix}^{-1},$$

and $(A'\ B'\ C') \neq (0\ 0\ 0)$ because M is non-singular and $(A\ B\ C) \neq (0\ 0\ 0)$.

Since in the projective plane there are no distinguished points, one expects that any point P can be transformed into another point P' by means of a projectivity; if that is so, the analogous property holds for lines, because of duality. In Exercise 4 the reader must prove these assertions, and that is why we say that the projective group is **transitive** on points and lines.

Let us now verify that the projective group is more extensive than the affine group. It will suffice to prove that a perspectivity can be given as an element of $PGL(3, \mathbb{R})$, but that it is not an affine transformation. Let us begin with the latter.

Proposition. *A perspectivity is* **not** *an affine transformation.*

3.5. The projective group

Proof. Let us consider Figure 3.14, which illustrates the affine plane, where the lines $\mathcal{L}$ and $\mathcal{M}$ correspond to the axes $y = 0$ and $x = 0$, and the point O from which we shall project $\mathcal{L}$ into $\mathcal{M}$ has coordinates $(1, 1, 1)$. The coordinate system is completed with the axis $z = 0$, which in the affine plane corresponds to the line at infinity.

The drawing shows that under the projection of $\mathcal{L}$ into $\mathcal{M}$ from O, the point at infinity $(0 : 1 : 0)$ of the affine plane is the image of a finite point $(x, 0, 1)$ (actually, we know that it is the point $(1, 0, 1)$), and instead the point at infinity of the line $\mathcal{L}$ (the axis $y = 0$), $(1 : 0 : 0)$ is projected into a finite point $(0, y, 1)$ (in this case we also know the point: $(0, 1, 1)$).

Since the type of point is invariant under affine transformations, we have proven that a perspectivity cannot be an affine transformation. $\square$

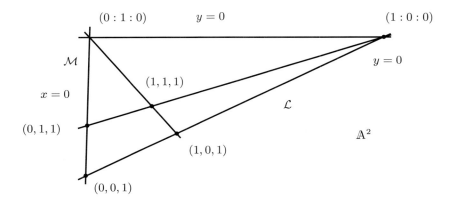

Figure 3.14: A perspectivity is not an affine transformation.

However, a perspectivity is certainly given by an element of the projective group:

Proposition. *A perspectivity is a projectivity.*

Proof. We must prove that given two lines $\mathcal{L}$ and $\mathcal{M}$, and a point exterior to both, O, there exists an element of $PGL(3, \mathbb{R})$ which maps the points P of $\mathcal{L}$ into the points P' of $\mathcal{M}$ so that P, O, and P' are collinear.

The adequate coordinate system is that where $\mathcal{L}$L corresponds to $y = 0$ and $\mathcal{M}$ corresponds to $x = 0$ (see Figure 3.15); the center of projection will be the point $O(1 : 1 : 1)$ and the third axis is a line different from the previous ones and which does not pass through O.

Notice that $(0 : y : 1)$ is collinear with $(x : 0 : 1)$ and $(1 : 1 : 1)$ if and only if $(x : 0 : 1)$ is collinear with $(0 : y : 1)$ and $(1 : 1 : 1)$; That is, a perspectivity is its own inverse (Exercise 2). The collinearity is equivalent to the nullity of the determinant formed with its coordinates:

$$\begin{vmatrix} x & 0 & 1 \\ 0 & y & 1 \\ 1 & 1 & 1 \end{vmatrix} = 0,$$

and computing the determinant we find that this happens when $y = x/(x-1)$.

Thus, our task is reduced to finding a matrix that maps the point $(x:0:1)$ into the point $(0:x:x-1)$ (Why?).

It is clear that the point common to $\mathcal{L}$ and $\mathcal{M}$, $(0:0:1)$, remains fixed, which means that (notice that on the right-hand side we only require that $\lambda \neq 0$):

$$\begin{pmatrix} a & b & c \\ d & e & f \\ g & h & i \end{pmatrix} \begin{pmatrix} 0 \\ 0 \\ 1 \end{pmatrix} = \begin{pmatrix} 0 \\ 0 \\ \lambda \end{pmatrix}, \quad \lambda \neq 0.$$

By doing the multiplication on the left-hand side we obtain $c = 0$, $f = 0$, and $i \neq 0$, and therefore the matrix has the form

$$\begin{pmatrix} a & b & 0 \\ d & e & 0 \\ g & h & i \end{pmatrix}, \quad i \neq 0.$$

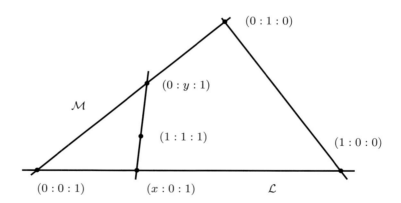

Figure 3.15: A convenient coordinate system for the proof.

By establishing the condition that the points of the line $\mathcal{L}$, $(x:0:z)$, be mapped into points of the line $\mathcal{M}$, $(0:y':z')$, we obtain $a = 0$ (verify it); this restricts the matrix even more:

$$\begin{pmatrix} 0 & b & 0 \\ d & e & 0 \\ g & h & i \end{pmatrix}, \quad i \neq 0.$$

3.5. The projective group

Let us now recall the remarks made with respect to Figure 3.14: the point $(1:0:1)$ is projected into the point $(0:j:0)$, with $j \neq 0$, and the point $(1:0:0)$ is projected into the point $(0:h:h)$, with $h \neq 0$. This implies that the matrix also satisfies that $d \neq 0$, $g + i = 0$, and $d = -i$, so it is of the form

$$\begin{pmatrix} 0 & b & 0 \\ -i & e & 0 \\ -i & h & i \end{pmatrix}.$$

The condition that defines the perspectivity implies that any line through $(1:1:1)$ transforms into itself, thus the point $(1:1:1)$ remains invariant:

$$\begin{pmatrix} 0 & b & 0 \\ -i & e & 0 \\ -i & h & i \end{pmatrix} \begin{pmatrix} 1 \\ 1 \\ 1 \end{pmatrix} = \begin{pmatrix} k \\ k \\ k \end{pmatrix}, \text{ for some } k \neq 0.$$

Multiplication on the left-hand side yields

$$\begin{pmatrix} b \\ -i + e \\ -i + h + i \end{pmatrix},$$

and since the three entries must be equal and different from zero, we have $b = -i + e = h$.

We now use the result of Exercise 2: a perspectivity is an involution. Hence, multiplying the matrix by itself we obtain the identity matrix (verify it!); the result gives $b = -i$; if we set $b = 1$, then the matrix is

$$\begin{pmatrix} 0 & 1 & 0 \\ 1 & 0 & 0 \\ 1 & 1 & -1 \end{pmatrix}.$$

That is why there is only one $T \in PGL(3, \mathbb{R})$ which gives the required perspectivity. □

Let us assume now that we have a perspectivity of $\mathcal{L}$ into $\mathcal{M}$ from O (see Figure 3.16), and another of $\mathcal{M}$ into $\mathcal{N}$ from O'; the composition is a projectivity, but it is not necessarily a perspectivity of $\mathcal{L}$ into $\mathcal{N}$, for the lines AA'', BB'' and CC'' are not necessarily concurrent.

But it is true that any projectivity between two lines, that is, a correspondence between their points established by an element of $PGL(3, \mathbb{R})$, can be expressed as the product of at most three perspectivities. This is a consequence of the so-called **Fundamental Theorem of Projective Geometry**, which we state and prove below.

Theorem. *Given two quadruples, A, B, C, D and A', B', C', D' of points of $P^2(\mathbb{R})$, each in general position, there exists one and only one projectivity $T \in PGL(3, \mathbb{R})$ such that $A' = T(A)$, $B' = T(B)$, $C' = T(C)$ and $D' = T(D)$.*

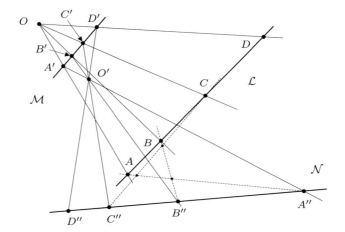

Figure 3.16: The projectivity that results from composing two perspectivities is not necessarily a perspectivity.

The proof is a consequence of the following Lemma.

Lemma. *Given four points $A, B, C, D \in P^2(\mathbb{R})$ in general position, it is possible to find representative elements for them such that $(a_1, a_2, a_3) = (b_1, b_2, b_3) + (c_1, c_2, c_3) + (d_1, d_2, d_3)$.*

Proof of the Lemma. Four projective points in general position are equivalent to four non-coplanar directions of $\mathbb{R}^3$. Thus, vectors with the direction of any three of them generate a basis for $\mathbb{R}^3$, so they determine a vector having the fourth direction:

$$(a_1, a_2, a_3) = \beta(b'_1, b'_2, b'_3) + \gamma(c'_1, c'_2, c'_3) + \delta(d'_1, d'_2, d'_3).$$

Therefore, we take $(b_1, b_2, b_3) = (\beta b'_1, \beta b'_2, \beta b'_3)$, $(c_1, c_2, c_3) = (\gamma c'_1, \gamma c'_2, \gamma c'_3)$, and $(d_1, d_2, d_3) = (\delta d'_1, \delta d'_2, \delta d'_3)$, to obtain the required representative elements. $\square$

Proof of the Theorem. We apply the lemma to each quadruple to obtain representative elements such that

$$\bar{A} = \bar{B} + \bar{C} + \bar{D}, \quad \bar{A}' = \bar{B}' + \bar{C}' + \bar{D}'.$$

Hence, if $M \in GL(3, \mathbb{R})$ is the only matrix that maps the basis $\bar{B}, \bar{C}, \bar{D}$ of $\mathbb{R}^3$ into the basis $\bar{B}', \bar{C}', \bar{D}'$, the linearity implies that $\bar{A}'^t = M\bar{A}^t$. $\square$

An immediate consequence is the following.

Corollary 1. *If $T \in PGL(3, \mathbb{R})$ leaves fixed four points in general position, then $T = \mathrm{id} \in PGL(3, \mathbb{R})$.*

3.5. The projective group

When the equivalent lemma and theorem are stated for the case of $P^1(\mathbb{R})$, it is easy to prove the following result, which refers to a range of points or to a pencil of lines illustrated in Figure 3.5.

Corollary 2. *A projective correspondence between two pencils of lines or two ranges of points is determined by two triples of corresponding points.*

As a consequence we have another corollary.

Corollary 3. *A projectivity between two pencils of lines or two ranges of points is the product of at most three perspectivities.*

Proof. Notice that if the point of intersection of two lines is mapped into itself, a projectivity between the two lines must be a perspectivity (Why?). Thus, given $A, B, C \in \mathcal{L}$ and $A', B', C' \in \mathcal{L}'$, the projectivity T between the lines $\mathcal{L}$ and $\mathcal{L}'$, which maps a triple into another, can be exhibited by introducing an auxiliary line $\mathcal{M}$ through A' (see Figure 3.17).

Now we project the points of $\mathcal{L}$ into $\mathcal{M}$ from $O = AA' \cap BB'$; the images of A, B and C are A', B'' and C''. Next we project the points of $\mathcal{M}$ into $\mathcal{L}'$ from $O' = B'B'' \cap C'C'''$: A' is mapped into itself, B'' into B', and C'' into C'. By composing both perspectivities we obtain the projectivity T.

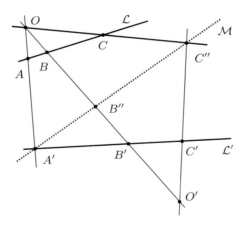

Figure 3.17: A projectivity between two different lines is a composition of two perspectivities.

When $\mathcal{L} = \mathcal{L}'$, the introduction of an auxiliary line $\mathcal{L}''$ into which we project the points of $\mathcal{L}'$ from O'' exterior to both lines, takes us to the previous situation (see Exercise 8). □

Before finishing this section, let us have a look on $P^1(\mathbb{C})$, to determine some properties of the transformations that act in it.

By analogy with the case of $P^2(\mathbb{R})$, the group whose invariants we are interested in studying must be $PGL(2,\mathbb{C})$, that is, classes of invertible 2×2-matrices whose entries a, b, c, d, are complex numbers and which act on an element of $(z:w) \in P^1(\mathbb{C})$ by multiplication:

$$\begin{pmatrix} a & b \\ c & d \end{pmatrix} \begin{pmatrix} z \\ w \end{pmatrix} = \begin{pmatrix} az+bw \\ cz+dw \end{pmatrix}.$$

Hence, except for the point of $P^1(\mathbb{C})$ with $w=0$, it makes sense (because $(z:1)$ is a representative element of a point with $w \neq 0$) to write

$$f(z) = \frac{az+b}{cz+d}, \qquad (3.12)$$

and the action of f can be extended to $\widehat{\mathbb{C}}$ following the usual rules of calculus when the variable z approaches infinity:

$$f(\infty) = a/c, \quad f(-d/c) = \infty.$$

Notice that in the first case the procedure is equivalent to using the chart corresponding to $z \neq 0$, that is, if the representative element is $(1:w)$, for $w=0$ we obtain the value a/c.

The expression (3.12) shows that we can restrict ourselves to matrices with determinant 1 (it suffices to divide the coefficients by the square root of the determinant of the matrix).

Therefore, the group that acts on $P^1(\mathbb{C})$ is $PSL(2,\mathbb{C})$, where S indicates that the determinant is 1. This group is better known as the group of **Möbius transformations**, and we are interested in it because it plays a relevant role in hyperbolic geometry, as we will see in Chapter 4.

For the moment we only prove that the Möbius transformations are **directly conformal** transformations of $P^1(\mathbb{C})$. This means that they respect angles including their orientations.

Proposition. *The Möbius transformations, $PSL(2,\mathbb{C})$, are directly conformal.*

Proof. It is enough to do the division indicated in (3.12):

$$f(z) = \frac{az+b}{cz+d} = \frac{a}{c} + \frac{b-(ad/c)}{cz+d},$$

since the right-hand side expresses f as a composition of directly conformal transformations (see Exercise 10)

$$z \mapsto cz \mapsto cz+d \mapsto \frac{1}{cz+d} \mapsto \frac{b-(ad/c)}{cz+d} \mapsto \frac{b-(ad/c)}{cz+d} + (a/c). \qquad \square$$

Exercises

1. Do the computations omitted in the proof that a perspectivity is an element of $PGL(3, \mathbb{R})$.

2. Prove that a perspectivity is an involution (i.e., it is its own inverse).

3. Determine the equation of the line into which the line defined by $2x - y + 4z = 0$ is transformed under the projectivity given by the matrix

$$\begin{pmatrix} 0 & 1 & 0 \\ -1 & -1 & -1 \\ 1 & 1 & -1 \end{pmatrix}.$$

4. Is it always possible to find a projectivity that maps a given point into another? What about any two lines? What happens if additionally we fix a point in each line? Justify your answers.

5. Prove that the image of a non-singular conic under a projectivity is another non-singular conic.

6. Find a projection in $\mathbb{R}^3$ that maps a circle into a hyperbola.

7. Prove Corollary 2.

8. Prove Corollary 3 when the two triples of points belong to the same line.

9. State and prove the Fundamental Theorem of Projective Geometry for $P^n(\mathbb{R})$.

10. Compare the Fundamental Theorem of Projective Geometry with the statement that the drawing of a tile determines the other tiles.

11. Verify that each of the types of transformations into which a Möbius transformation is decomposed is conformal.

3.6 Invariance of the cross ratio

Once we have proved that $PGL(3, \mathbb{R})$ is the group of projectivities, our task now is to find the associated invariants. One of them, fundamental for a great part of the theory, is numeric and it refers to the cross ratio of four elements of a pencil or a range, which we shall think of as $P^1(\mathbb{R})$.

Let us start by observing that a perspectivity does not respect the ratio into which a point divides a segment. Figure 3.18 shows an isosceles triangle, and when its base AB is projected from C into a line through B, different from the base, the mid-point M is mapped into a point M' which does not divide the segment $A'B$ into equal parts.

Thus, the ratio into which a point divides a segment is not a projective invariant; this is why it may seem surprising that the **cross ratio**, $(A, B; C, D)$, among four collinear points, A, B, C, and D, is a projective invariant. The cross ratio for points of the real line has the form

$$(A, B; C, D) = \frac{\frac{C-A}{B-C}}{\frac{D-A}{B-D}} = \frac{(A-C)(D-B)}{(C-B)(A-D)}. \qquad (3.13)$$

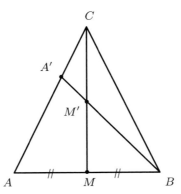

Figure 3.18: The ratio into which a point divides a segment is not a projective invariant.

The double quotient (3.13) is formed, as exhibited by the middle member, as the quotient of two ratios: the ratio into which C divides the segment AB by the ratio into which D divides that same segment.

Notice that the order in which the points are written is very important. To convince oneself of that, the classic Exercise 4 proves that given four fixed collinear points, there are only six possible values for the cross ratio, depending on the order in which the points are written. For instance, it is very easy to verify that if we exchange the pairs, keeping the order in them, the cross ratio is preserved:

$$(A, B; C, D) = (C, D; A, B).$$

In Figure 3.19 we have marked the angles α, β, γ and δ; it is left to the reader as an exercise (Exercise 3) to prove, using the Law of Sines, that $(A, B; C, D)$ and $(A', B'; C', D')$ have the same expression in terms of the sines of those angles:

$$\frac{\sin \alpha \, \sin \delta}{\sin \beta \, \sin \gamma}.$$

One particularly interesting case occurs when the value of the cross ratio $(A, B; C, D)$ is -1; it is said then that A, B, C, D form a **harmonic set** of points (as the reader may suspect, the term comes from music), and that C and D are **harmonic conjugates** one to another with respect to A, B.[1]

---

[1] If a chord pressed at the points A and B gives a certain note, by pressing it at the points C and D, the other notes of the major triad are obtained.

3.6. Invariance of the cross ratio

In the study of projective invariants, there are several sets of harmonic points that arise naturally, as will be seen below.

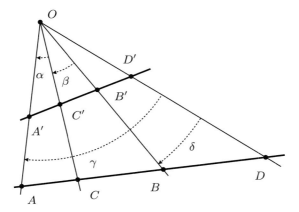

Figure 3.19: The cross ratio $(A, B; C, D)$ is equal to the cross ratio $(A', B'; C', D')$.

The way of extending the definition of **cross ratio** to four points of a projective line, with coordinates $(a_1 : a_2)$, $(b_1 : b_2)$, $(c_1 : c_2)$, $(d_1 : d_2)$, is

$$(A, B; C, D) = \frac{\begin{vmatrix} a_1 & c_1 \\ a_2 & c_2 \end{vmatrix} \begin{vmatrix} d_1 & b_1 \\ d_2 & b_2 \end{vmatrix}}{\begin{vmatrix} c_1 & b_1 \\ c_2 & b_2 \end{vmatrix} \begin{vmatrix} a_1 & d_1 \\ a_2 & d_2 \end{vmatrix}}. \tag{3.14}$$

Verifying that the definition does not depend on the representative elements chosen for the points is immediate, because of the properties of determinants. And to prove that this definition extends that of the Euclidean case, compute the value of the cross ratio when the second coordinate of each point is 1.

In the Exercises section there are several expressions for the cross ratio that it is worthwhile to know, so that we can use the most adequate of them at a given situation.

To prove the invariance of the cross ratio among four elements of a pencil or a range under projectivities, we first consider that by eliminating the absolute value in the numerator of formula (1.2), we obtain the **oriented distance** from a point $P_0(x_0, y_0) \in \mathbb{R}^2$ to the line $\mathcal{L}$ of equation $Ax + By + C = 0$:

$$d(P_0, \mathcal{L}) = \frac{Ax_0 + By_0 + C}{\sqrt{A^2 + B^2}}. \tag{3.15}$$

Let us now consider two lines $\mathcal{A}$ and $\mathcal{B}$, with equations

$$\alpha(x, y) = A_1 x + B_1 y + C_1 = 0 \text{ and } \beta(x, y) = A_2 x + B_2 y + C_2 = 0,$$

which we represent by $\alpha = (A_1, B_1, C_1)$ and $\beta = (A_2, B_2, C_2)$.

With that notation, the set of points $P(x, y) \in \mathbb{R}^2$ whose distances to $\mathcal{A}$ and $\mathcal{B}$ are in the constant ratio k is the line $\mathcal{C}$ corresponding to $\gamma(x, y) = 0$, where the coefficients of γ are given by

$$\gamma = \alpha - k \frac{\sqrt{A_1^2 + B_1^2}}{\sqrt{A_2^2 + B_2^2}} \beta, \qquad (3.16)$$

as can be verified easily by writing the two distances in the form (3.15) and requiring their quotient to be k.

The interesting geometric interpretation of the above discussion is that the line $\mathcal{C}$ represented by (3.16) divides the pair $(\mathcal{A}, \mathcal{B})$ in the ratio k, and if we have another line $\mathcal{D}$ with equation $\delta(x, y) = 0$ given by

$$\delta(x, y) = \alpha - h \frac{\sqrt{A_1^2 + B_1^2}}{\sqrt{A_2^2 + B_2^2}} \beta, \qquad (3.17)$$

the cross ratio of the lines $(\mathcal{A}, \mathcal{B}; \mathcal{C}, \mathcal{D})$ is simply k/h.

Let us write this result as a lemma, with the simplified notation that takes into account that, for the projective lines obtained by homogenizing the equations, the coefficients of the combination provide coordinates for the lines of the pencil defined by $(\mathcal{A}, \mathcal{B})$.

Lemma 1. *If the concurrent lines $\mathcal{A}$, $\mathcal{B}$, $\mathcal{C}$ and $\mathcal{D}$ have coordinates α, β, $\alpha + \lambda\beta$ and $\alpha + \mu\beta$, then the value of the cross ratio $(\mathcal{A}, \mathcal{B}; \mathcal{C}, \mathcal{D})$ is*

$$(\mathcal{A}, \mathcal{B}; \mathcal{C}, \mathcal{D}) = \frac{\lambda}{\mu}.$$

Notice that it is not necessary to normalize the equations, for the factor $\frac{\sqrt{A_1^2 + B_1^2}}{\sqrt{A_2^2 + B_2^2}}$ would appear in the numerator as well as in the denominator.

To prove the invariance under $T \in PGL(3, \mathbb{R})$ of the cross ratio of four collinear points A, B, C and D, we just need to establish a second lemma.

Lemma 2. *If A, B, C and D are four collinear points with coordinates*

$$(a_1 : a_2 : a_3); (b_1 : b_2 : b_3) \quad ;$$
$$(a_1 : a_2 : a_3) + \lambda(b_1 : b_2 : b_3) \quad ; \quad (a_1 : a_2 : a_3) + \mu(b_1 : b_2 : b_3),$$

respectively, by joining them to a point $(r_1 : r_2 : r_3)$ exterior to their line, we obtain lines $\mathcal{A}$, $\mathcal{B}$, $\mathcal{C}$ and $\mathcal{D}$ whose cross ratio is

$$(\mathcal{A}, \mathcal{B}; \mathcal{C}, \mathcal{D}) = \frac{\lambda}{\mu}.$$

3.6. Invariance of the cross ratio

Proof. It suffices to write the equations of the lines by means of the nullity of the determinant whose columns are $(x, y, z)^t$, $(a_1, a_2, a_3)^t$, and $(r_1, r_2, r_3)^t$, for $\alpha(x, y, z) = 0$, and analogously for the other three. In the case of $\mathcal{C}$ and $\mathcal{D}$, the middle column is a combination of two columns and, by the properties of determinants, the equations of the four lines have precisely the form given in Lemma 1, proving this second lemma. $\square$

With these results, the invariance of the cross ratio under projectivities is immediate.

Theorem. *The cross ratio $(A, B; C, D)$ of four collinear points A, B, C, D is invariant under any $T \in PGL(3, \mathbb{R})$.*

Proof. The collinear points A, B, C, D are mapped under T into collinear points A', B', C', D'. The argument used in the proof of the lemma of Section 3.5, insures that it is possible to find representative elements of the first three points such that their coordinates are

$$(a_1 : a_2 : a_3), \quad (b_1 : b_2 : b_3), \quad (a_1 : a_2 : a_3) + (b_1 : b_2 : b_3),$$

and then the linear combination of $(a_1 : a_2 : a_3)$ and $(b_1 : b_2 : b_3)$, which gives the coordinates of D, is determined by the value λ of the cross ratio $(A, B; C, D)$: $(a_1 : a_2 : a_3) + \lambda^{-1}(b_1 : b_2 : b_3)$.

By applying T to each of those points, the combinations of the coordinates of A' and B' corresponding to C' and D' coincide with those of C and D; accordingly, $(A, B; C, D) = (A', B'; C', D')$. $\square$

Exercises

1. Consider the points -2, 0, 3 and 5 of the real line, and establish all the possible cross ratios among them.

2. Prove that for points in $\mathbb{R}^2$, the formulas for the coordinates of the point $P(x, y)$ which divides the segment $P_1 P_2$ in the ratio μ are

$$x = \frac{x_1 - \mu x_2}{1 - \mu}, \quad y = \frac{y_1 - \mu y_2}{1 - \mu}.$$

   (Suggestion: Consider the similar triangles that are formed by drawing through P_1 the line parallel to the X axis, and through P and P_2 the lines parallel to the Y axis.)

3. In Figure 3.19, prove that $(A, B; C, D) = (A', B'; C', D')$ by expressing both cross ratios in terms of the sines of the angles α, β, γ and δ.

4. Prove that given four fixed collinear points, the only possible values for the cross ratios that can be established among them are $\lambda, 1 - \lambda, \lambda/(1 - \lambda)$ and their reciprocals.

5. Prove that four points of a range form a harmonic set if and only if $(A, B; C, D) = (A, B; D, C)$.

6. State and prove results for lines of a pencil, analogous to those proposed in the previous exercises.

7. In each case, prove that the cross ratio has the proposed value when the elements of the pencil of concurrent lines or a range of collinear points have the expressions given below (we write $\bar{a}$ for $(a_1 : a_2)$):

$$(\bar{a}, \bar{b}; k_1\bar{a} + h_1\bar{b}, k_2\bar{a} + h_2\bar{b}) = \frac{h_1 k_2}{h_2 k_1};$$

$$(\bar{a} + k_1\bar{b}, \bar{a} + k_2\bar{b}; \bar{a} + k_3\bar{b}, \bar{a} + k_4\bar{b}) = \frac{(k_3 - k_1)(k_4 - k_2)}{(k_3 - k_2)(k_4 - k_1)};$$

$$(k_1\bar{a} + h_1\bar{b}, k_2\bar{a} + h_2\bar{b}; k_3\bar{a} + h_3\bar{b}, k_4\bar{a} + h_4\bar{b}) = \frac{\begin{vmatrix} k_3 & h_3 \\ k_1 & h_1 \end{vmatrix} \begin{vmatrix} k_4 & h_4 \\ k_2 & h_2 \end{vmatrix}}{\begin{vmatrix} k_3 & h_3 \\ k_2 & h_2 \end{vmatrix} \begin{vmatrix} k_4 & h_4 \\ k_1 & h_1 \end{vmatrix}}.$$

3.7 The space of conics

At the beginning of this chapter we presented a drawing that intuitively justifies the fact that

All non-singular conics are projectively equivalent.

Now we are ready to prove this statement formally. This follows by observing, as we show below, that to each non-singular conic there corresponds a non-singular symmetric matrix M (rather its class), that is, an element of $PGL(3, \mathbb{R})$, and conversely. Since $PGL(3, \mathbb{R})$ is a group, for any two elements M and M' there exists another element, T, thought of as a projective transformation such that $M' = TM$. Hence all non-singular conics are projectively equivalent.

The way of assigning a symmetric matrix (actually a class) to a projective conic comes from writing the following equation as a product of matrices:

$$Ax^2 + 2Bxy + Cy^2 + 2Dxz + 2Eyz + Fz^2 = 0,$$

obtained by homogenizing the Cartesian equation of a conic, $Ax^2 + 2Bxy + Cy^2 + 2Dx + 2Ey + F = 0$.

The reader can verify that the first member of the previous equation is

$$\begin{pmatrix} x & y & z \end{pmatrix} \begin{pmatrix} A & B & D \\ B & C & E \\ D & E & F \end{pmatrix} \begin{pmatrix} x \\ y \\ z \end{pmatrix}. \tag{3.18}$$

In addition it is also easy to prove the uniqueness of the class of the symmetric matrix, which henceforth will be called the matrix of the conic.

3.7. The space of conics

By using those matrices, the classification of the projective conics is simple, for we can diagonalize the matrix because it is symmetric. Furthermore, by using an adequate projectivity, the entries of the diagonal are 1 or -1. Hence, if the conic is non-singular, then its rank is 3, and for it to be non-empty, its signature must be 1. Therefore its standard form is

$$x^2 + y^2 - z^2 = 0.$$

In the singular case there are several possibilities and singular conics do not have a unique standard form. Since their corresponding matrix must be singular, their rank is 2 or 1.

When the rank is 2, the signature can be 2 or 0, and the standard forms are the following:

- $x^2 + y^2 = 0$, which corresponds to the projective point $(0:0:z)$;
- $x^2 - y^2 = 0$, which corresponds to two projective lines, $(x:x:z)$ and $(x:-x:z)$.

Finally, when the rank is 1, the standard form is

- $x^2 = 0$, whose locus is the double line $(0:y:z)$.

Let us now analyze the conics from another point of view.

Since to each projective conic there corresponds a symmetric 3×3-matrix, we can parametrize the space of conics with the entries of this type of matrices: three parameters in the diagonal and three above the diagonal (a total of six). However, by forming classes we lose one dimension, and that is why we essentially have five parameters, that corresponds to the geometric fact that fivepoints, and not fewer, determine a conic.

That is, each projective conic determines an element of $P^5(\mathbb{R})$ and vice-versa, although the mapping is not bijective, since any point in $P^5(\mathbb{R})$ whose coordinates have all the same sign corresponds to the empty set. That would not happen if the coordinates were complex numbers, for in that case the locus of an equation such as $x^2 + y^2 + z^2 = 0$ is not the empty set.

To avoid the situation when an equation does not have roots, in algebraic geometry the variables take values in **algebraically closed fields**, where any polynomial with coefficients in it has a root and, accordingly, all the roots are in the field. The Fundamental Theorem of Algebra, due to Karl Friedrich Gauss (1777–1855), establishes that $\mathbb{C}$ has that property (see [B-ML] or [A]).

For the rest of this section, we shall assume that the variables take values in $\mathbb{C}$; the classification of the conics then is simplified because the only invariant is the rank (we leave this as an exercise for the reader).

If we take a random element $(A:B:C:D:E:F)$ of $P^5(\mathbb{C})$, it is most likely that it determines a non-singular conic, since that is equivalent with asking that the determinant of the symmetric matrix is not zero.

When the determinant is zero, the coordinates of the point $(A : B : C : D : E : F) \in P^5(\mathbb{C})$ must satisfy the third-degree equation

$$ACF + 2BED - CD^2 - AE^2 - FB^2 = 0. \tag{3.19}$$

This equation defines a hypersurface in $P^5(\mathbb{C})$, whose complement is an open set formed by points representing non-singular conics.

The term **hypersurface** only indicates that one degree of freedom has been lost, which is due to the restriction imposed by the equation; in this case, we subtract 1 from the dimension 5, and therefore we say that the conics of rank ≤ 2 determine a subset of dimension 4 in $P^5(\mathbb{C})$.

We can also obtain that conclusion about the dimension by observing that a conic of rank 2 represents two different lines (complex lines), and since in $P^2(\mathbb{C})$ duality also holds, the space of lines has dimension 2. Accordingly, the dimension of the space of pairs of lines is 4.

When the rank is 1, all the subdeterminants of order 2 in the matrix of the conic become zero (which implies that equation (3.19) is satisfied). That might seem to establish too many equations, each would make us lose one degree of freedom. However, if the readers write the determinants, they will find out that only three of the determined equations are linearly independent, that is why the dimension of the space of conics of rank 1 is 2. This space is contained (as a closed subset) in the hypersurface defined by (3.19); its complement in this hypersurface consists of the conics of rank 2.

Of course, the conclusion about the dimension is immediate if we take into account that each conic of rank 1 determines one projective (double) line, and that the dimension of the space of the lines in $P^2(\mathbb{C})$ is 2.

Before concluding this section, we introduce the **Veronese mapping**, which can be defined both over $\mathbb{R}$ and over $\mathbb{C}$:

$$v : P^2(\mathbb{R}) \to P^5(\mathbb{R})$$

given by $(x : y : z) \mapsto (x^2 : y^2 : z^2 : 2xy : 2xz : 2yz)$. This mapping is clearly differentiable and injective (verify it), and that is why the image has dimension 2. Hence its image is a surface in $P^5(\mathbb{R})$, called the **Veronese surface**.

Under this mapping, each conic $\mathcal{C} \subset P^2(\mathbb{R})$ belongs to a hyperplane of $P^5(\mathbb{R})$, whose equation has as coefficients the entries of the matrix of the conic

$$Ax_1 + Bx_2 + Cx_3 + Dx_4 + Ex_5 + Fx_6 = 0.$$

The dual space of $P^5(\mathbb{R})$ consists precisely of all hyperplanes, and the correspondence between conics of $P^2(\mathbb{R})$ and hyperplanes of $P^5(\mathbb{R})$ can be interpreted as a correspondence between the conics of $P^2(\mathbb{R})$ and the dual space $(P^5(\mathbb{R}))^*$ of $P^5(\mathbb{R})$.

Exercises

1. Verify that by projectivizing the equation of a non-singular conic, the associated matrix is non-singular.

2. In $\mathbb{R}^2$, give the equations of two conics passing through the same four points. Use those equations to give two different conics in $P^2(\mathbb{R})$ intersecting at four points.

3. Classify the conics in $P^2(\mathbb{C})$. (Suggestion: Remember that in this case the only invariant is the rank.)

4. Prove that the Veronese mapping is injective.

3.8 Projective properties of the conics

In Figure 3.20 we have drawn an ellipse in which we have fixed six arbitrary points, numbered from 1 to 6; the line 12 intersects the line 45 at a point P, the line 23 intersects the line 56 at a point Q, and the line 34 intersects the line 61 at a point R. Now draw the line PQ; we can observe that it seems to pass through the point R. Surprisingly, one has that this is actually so!

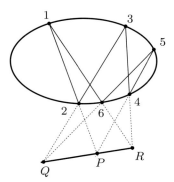

Figure 3.20: The intersections of opposite sides of an hexagon inscribed into a conic are colinear.

One of the purposes of this section is to prove that collinearity, which was discovered by Blaise Pascal (1623–1662) who, astonished, decided to call *mystic* a hexagon inscribed into a conic.

Notice that a particular case of Pascal's Theorem is the Theorem of Pappus (whose proof is left to the reader).

Theorem of Pappus. *If 1, 3 and 5 are three points of a line, and 2, 4 and 6 are three points of another line, then the intersection of 12 with 45, that of 23 with 56, and that of 34 with 61 are collinear.*

However, Pascal's theorem is probably not the most amazing result about conics. For instance, we have this other result: take four points A, B, C, D of a conic (see Figure 3.21), and consider the lines they form with two fixed points, O and O', also in the conic. There always exists a projectivity between the pencils defined by O and O' which maps OA, OB, OC and OD into $O'A$, $O'B$, $O'C$ and $O'D$.

The converse of this result is also true, and both are briefly stated as follows:

Theorem (Projective Characterization of the Conics). *Every conic is the locus of the intersections of corresponding lines of two pencils with vertices O and O', the correspondence being given by a projectivity.*

This result is actually a corollary of the following theorem of Jacob Steiner (1796–1863).

Steiner's Theorem. *Given four points A, B, C, D in a conic, the cross ratio of the lines they determine with a fifth point O in the conic does not depend on the point O.*

Proof. The proof reduces to verifying that if O' is another point in the conic, then

$$(OA, OB; OC, OD) = (O'A, O'B; O'C, O'D), \qquad (3.20)$$

where OA denotes the line of the pencil defined by O which contains A, $O'A$ denotes the line of the pencil defined by O' which contains A, and analogously for B, C, and D.

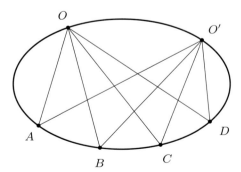

Figure 3.21: Being fixed $A, B, C, D \in \mathcal{C}$, for any $O, O' \in \mathcal{C}$, it occurs that $(OA, OB; OC, OD) = (O'A, O'B; O'C, O'D)$.

Let us prove that the above equation is always satisfied.

3.8. Projective properties of the conics

We know that we can find coordinates α and β for OA and OB such that $\alpha + \beta$ are the coordinates of OC. Therefore, by (3.16), the cross ratio of the lines through O reduces to the quotient $-\beta(D)/\alpha(D)$.

We can also find representative elements for the lines through O' such that if α' and β' are the coordinates of $O'A$ and $O'B$, then $\alpha' + \beta'$ are the coordinates of $O'C$, and for the same argument given before, the cross ratio of the lines through O' is $-\beta'(D)/\alpha'(D)$.

Hence, proving (3.20) reduces to verifying the equality

$$\beta'(D)/\alpha'(D) = \beta(D)/\alpha(D). \tag{3.21}$$

For this we write the equations of the pairs of lines OA and $O'B$, and $O'A$ and OB, which are respectively,

$$(\alpha \bar{x}^t)(\beta' \bar{x}^t) = 0 \quad \text{and} \quad (\alpha' \bar{x}^t)(\beta \bar{x}^t) = 0 \,.$$

Since the four points O, A, O', B also belong to the conic $\mathcal{C}$, its equation must be a linear combination of the two previous equations:

$$r(\alpha \bar{x}^t)(\beta' \bar{x}^t) + s(\alpha' \bar{x}^t)(\beta \bar{x}^t) = 0.$$

The same can be said with respect to the pairs of lines OA, $O'C$ and $O'A$, OC; thus, the equation of $\mathcal{C}$ is also a linear combination of the two equations

$$(\alpha \bar{x}^t)(\alpha' \bar{x}^t + \beta' \bar{x}^t) = 0 \quad \text{and} \quad (\alpha' \bar{x}^t)(\alpha \bar{x}^t + \beta \bar{x}^t) = 0.$$

If the coefficients of this other linear combination are ρ and σ, we have

$$\rho \alpha \alpha' + \rho \alpha \beta' + \sigma \alpha \alpha' + \sigma \alpha' \beta = 0.$$

Since B renders the second and the fourth terms, but does not satisfy $\alpha \alpha' = 0$, the coefficient $\rho + \sigma$ of the $\alpha \alpha'$ term must be zero; that implies that $r = -s = 1$. Therefore, the equation of $\mathcal{C}$ is

$$(\alpha \bar{x}^t)(\beta' \bar{x}^t) - (\alpha' \bar{x}^t)(\beta \bar{x}^t) = 0,$$

which we obtain the equality (3.21). □

Equation (3.20) implies that it makes sense to define the **cross ratio of four points in a conic** precisely as the cross ratio of the lines of the pencil with vertex $O \in \mathcal{C}$ containing those four points, since that number does not depend on the choice of $O \in \mathcal{C}$.

The proof of the Theorem of the Projective Characterization of the Conics is left now to the reader (Exercise 1). Pascal's Theorem is a simple consequence of Steiner's Theorem, as will be seen below.

Theorem (of the Mystic Hexagon). *In any hexagon inscribed into a conic, the intersections of opposite sides are collinear.*

Proof. We decided to denote the points with numbers because then the opposite sides (obtained by "skipping" two consecutive sides) are obtained by adding 3 modulo 6 to the vertices of the elected side: 12 and 45 with intersection P, 23 and 56 with intersection Q, and 34 and 61 with intersection R.

Figure 3.22, which reproduces Figure 3.20, not only shows the intersections P, Q and R of opposite sides, but also $S = 12 \cap 56$ and $T = 16 \cap 45$, which will be useful for establishing cross ratios in the lines 56 and 16, whose equality will be established by means of projections into the conic and by Steiner's Theorem.

What has to be proven — the collinearity of P, Q and R — will follow from the fact that the equality of cross ratios of lines in two pencils implies that $PQ = PR$.

We write $(Q, S; 6, 5)$ for the cross ratio of the four points on the side 56 of the hexagon. If we project these points into the conic from the vertex 2, we obtain, respectively, 3, 1, 6 and 5; therefore,

$$(Q, S; 6, 5) = (3, 1; 6, 5).$$

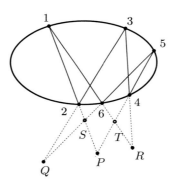

Figure 3.22: P, Q, and R are colinear.

If now we project these points of the conic into the line 16 from 4, we obtain, respectively, R, 1, 6 and T; thus,

$$(Q, S; 6, 5) = (3, 1; 6, 5) = (R, 1; 6, T).$$

Hence, there is a projectivity of the pencil with vertex at P onto itself such that

$$(PQ, PS; P6, P5) = (PR, P1; P6, PT).$$

The lines PS and $P1$ are the same, and also the lines $P5$ and PT are the same. Therefore, the lines PQ and PR are the same as well, as was stated. □

Before finishing this section, we shall establish a very important property of a quadrilateral, which generalizes the following fact about any Euclidean parallelogram (see Figure 3.23).

3.8. Projective properties of the conics

A line parallel to two sides of a parallelogram through the point of intersection of its diagonals intersects both of the other sides at their mid-points.

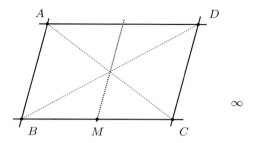

Figure 3.23: M is the mid-point of BC.

Notice that this is equivalent to saying that $(B, C; M, \infty)$ form a harmonic set of points, where ∞ is the point at infinity of the line BC.

Since in projective geometry there are no parallel lines, the statement takes the following form.

Lemma. *Let A, B, C, D, F and G be the vertices of a quadrilateral, and take the three vertices B, C, and F belonging to one of its sides. The pairs of vertices A, C and B, D are opposite vertices, and the lines AC and BD intersect in a point L such that the line LG meets the first side in a point M. The points B, C, M, and F form a harmonic set of points, that is, $(B, C; M, F) = -1$ (see Figure 3.24).*

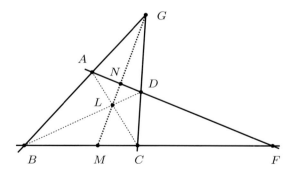

Figure 3.24: B, C, M and F form a harmonic set.

Proof. It was left to the reader to prove that $(B, C; M, F) = (C, B; M, F)^{-1}$; thus B, C, M and F form a harmonic set of points if and only if $(B, C; M, F) = (C, B; M, F)$.

Again, it is enough to project conveniently to obtain the wanted equality: first we project from L the side BC into the side AD, and then we project from

G the side AD into the side BC,
$$(B, C; M, F) = (D, A; N, F) = (C, B; M, F).$$
The equality of the first and last members is what we wanted. □

Exercises

1. Prove that a conic is the locus of the intersections of corresponding lines of two pencils related by a projectivity.
2. Prove Pascal's Theorem in the case of the hexagon shown below.

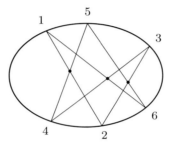

3. Dualize the concept of conic using the projective characterization of conic.
4. Dualize Pascal's Theorem; this proposition is called Brianchon's Theorem.
5. Prove the Theorem of Pappus.
6. Analyze what happens with Pascal's Theorem when two of the vertices are mixed, i.e., when one "side" becomes tangent to the conic.

3.9 Poles and polars

When we dealt with the theme of duality, we used the concept of orthogonality in $\mathbb{R}^3$ several times, where the scalar product is the usual one.

Any scalar product in $\mathbb{R}^3$ gives rise to a symmetric matrix (it is called the **matrix of the metric**) (see [B-ML] or [Ri]), which in the case of the usual scalar product is the identity matrix (for the product of matrices), and that is why we omit it when writing $(a, b, c) \cdot (r, s, t)$; in the general case, to obtain the scalar product of any two vectors, we would write

$$\begin{pmatrix} a & b & c \end{pmatrix} \begin{pmatrix} A & B & D \\ B & C & E \\ D & E & F \end{pmatrix} \begin{pmatrix} r \\ s \\ t \end{pmatrix}.$$

3.9. Poles and polars

The symmetric matrix can be thought of as the matrix of a conic $\mathcal{C}$,

$$\begin{pmatrix} x & y & z \end{pmatrix} \begin{pmatrix} A & B & D \\ B & C & E \\ D & E & F \end{pmatrix} \begin{pmatrix} x \\ y \\ z \end{pmatrix} = 0, \qquad (3.22)$$

and when $P_0(x_0 : y_0 : z_0)$ does not belong to the conic, algebra shows that the locus of the points $P(x : y : z)$ that satisfy the equation

$$\begin{pmatrix} x_0 & y_0 & z_0 \end{pmatrix} \begin{pmatrix} A & B & D \\ B & C & E \\ D & E & F \end{pmatrix} \begin{pmatrix} x \\ y \\ z \end{pmatrix} = 0, \qquad (3.23)$$

is a straight line, called the **polar line of** P_0 with respect to $\mathcal{C}$, and any point in that line is a **polar point** of P_0 (see Figure 3.26).

Notice that the symmetry of the matrix implies that the relation "A is polar of B with respect to $\mathcal{C}$" is a symmetric relation.

Notice as well that the points in the conic are polars of themselves, and that the polars of collinear points are concurrent lines (prove it).

From our discussion about the form of $P^2(\mathbb{R})$, the reader will agree that there is an essential difference between a projective line and a conic, in spite of both being topological circles: the first one does not separate $P^2(\mathbb{R})$ into two disjoint regions (the middle line of the strip is a projective line, for it comes from a circle of maximal radius), while a conic does (the edge of the strip is a conic, and since a conic comes from two antipodal parallels, it satisfies a second degree equation). Of the two regions mentioned above, one is homeomorphic to a disk and the other to a Möbius strip (see Figure 3.25).

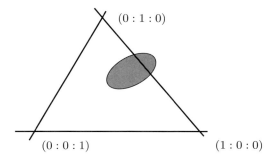

Figure 3.25: A conic separates $P^2(\mathbb{R})$ into two non-homeomorphic parts.

Thus, a point P_0 which does not belong to the conic may lie inside the conic, and in such a case there are no tangent lines to the conic through P_0 (prove it algebraically); however, if the point lies outside the conic, in the region homeomorphic to the strip, there are two lines tangent to the conic passing through P_0, and the points of tangency determine the polar.

We understand as a **line tangent** to $\mathcal{C}$ precisely a line which intersects $\mathcal{C}$ at a double point, that is, if the line is determined by two points R and S, any point of the line different from S is of the form $R + \lambda S$, and by substituting this point in equation (3.23), we obtain a second-degree equation in λ:

$$(R + \lambda S)M(R + \lambda S)^t = 0,$$

$$(R + \lambda S)M(R + \lambda S)^t = 0,$$

where M denotes the matrix of the conic.

If we develop this equation, we obtain

$$\lambda^2 (SMS^t) + 2\lambda(SMR^t) + RMR^t = 0,$$

and the root is double when the discriminant becomes zero:

$$(SMR^t)^2 - (SMS^t)(RMR^t) = 0. \tag{3.24}$$

Now it is easy to determine the polar of a point R exterior to the conic (see Figure 3.26).

Proposition. *If it is possible to draw two tangent lines from R to a conic, and the points of tangency are S_1 and S_2, then the polar of R is the line S_1S_2.*

Proof. It will suffice to prove that both S_1 and S_2 belong to the polar of R. Since S_1 belongs to the conic, the equation (3.24) reduces to $S_1MR^t = 0$, which shows that S_1 belongs to the polar of R.

The same holds for S_2, and therefore every linear combination of both also belongs to the polar. $\square$

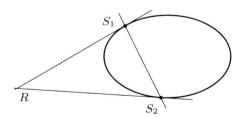

Figure 3.26: The points of tangency determine the polar.

If the point R lies inside the conic, its polar can be drawn by applying the proposition we have just proven in the following way. Take two lines $\mathcal{L}$ and $\mathcal{L}'$ through R; each intersects the conic at two distinct points (prove it), A and A' for L, and B and B' for L' (see Figure 3.27).

3.9. Poles and polars

The tangent lines to $\mathcal{C}$ at A and A' intersect at a point X, and according to the construction given in the last proposition, the polar of X is the line through A and A'. Analogously for B and B': the tangent lines to $\mathcal{L}$ through B and B' intersect at a point Y whose polar is precisely the line through B and B'.

Hence X and Y are polars with respect to R, which implies that the line through X and Y is the polar of R.

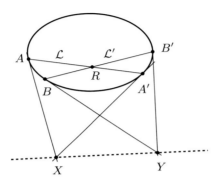

Figure 3.27: Construction of the polar for a point R in the interior of $\mathcal{C}$

The concept of polarity can be used to solve an important problem of linear algebra, which is the simultaneous diagonalization of two quadric forms.

For this theme we follow one of our main references, [Re].

Definition. A triangle ABC is called a **self-polar triangle** with respect to a conic C if each side is the polar of the opposite vertex.

The condition might seem very restrictive, and it is reasonable to ask whether for any conic there exist self-polar triangles.

The answer is positive, as can be easily verified (make a drawing): If a point A exterior to $\mathcal{C}$ is fixed, and we seek BC to be its polar with respect to $\mathcal{C}$, we can choose B arbitrarily in the polar of A, and then C, by definition of a self-polar triangle, must be the intersection of the polar of A with the polar of B.

Let us now see what is the equation of the conic $\mathcal{C}$ when the triangle of reference is a self-polar triangle with respect to the conic; the sides of the triangle have equations $x = 0$, $y = 0$, and $z = 0$, and each of the points of intersection of the axes is polar of the other two.

The coordinates of those points are $(0:0:1)$, $(0:1:0)$ and $(1:0:0)$, and the condition of polarity implies that the matrix of the conic is diagonal.

Even in basic courses of analytic geometry one learns about the advantages of the matrix of a conic (or quadric) being diagonalizable: under that form (the standard form), it is very easy to recognize the conic. Later on, the need of diagonalizing two quadric forms simultaneously arises.

That is why it would be very useful that there existed self-polar triangles with respect to two conics, for by taking one such triangle as reference, the matrices of both conics would have diagonal form.

The important question now follows:

Given two conics, is there a triangle being self-polar with respect to both?

A question like this should always be analyzed from the point of view of the degrees of freedom, that is, of the dimensions of the objects involved.

The space of the triangles in $P^2(\mathbb{R})$ has dimension 6 (because each vertex can be taken arbitrarily in the projective plane), and the dimension of the space of the self-polar triangles with respect to a conic has dimension 3, since the first vertex can be taken arbitrarily in a space of dimension 2, $P^2(\mathbb{R})$, the second vertex can be taken arbitrarily in a space of dimension 1, the polar of the first vertex, and there are no degrees of freedom for the last vertex.

If now we have two conics, claiming that there exist self-polar triangles with respect to both conics is asking for two three-dimensional spaces of a space of dimension 6, the space of the triangles, to intersect.

Broadly speaking, a way of determining a three-dimensional space in a space of dimension 6 consists in establishing three linearly independent polynomial equations, because each equation eliminates one degree of freedom; in addition, for two three-dimensional spaces to have common points, we need a system of six equations with six unknowns to be solvable.

Stated in these terms, the problem seems to have a solution, although we must remember that not every polynomial equation of degree greater than 1 with real coefficients has a real solution. That is why it is necessary to establish a condition which guarantees that we have solutions. This condition is not too restrictive (see Figure 3.29).

Theorem. *Given two conics which intersect at four points, there always exists a self-polar triangle with respect to both conics.*

Before proving the theorem, we prove a very interesting property of the points of intersection of a conic and a line.

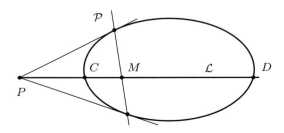

Figure 3.28: P, M, C and D, form a harmonic set.

3.9. Poles and polars

Lemma. *Given a point P which does not belong to a conic $\mathcal{C}$ and a line through P that intersects $\mathcal{C}$ at C and D, if M is the intersection of the polar $\mathcal{P}$ of P, then $(P, M; C, D) = -1$. Conversely, if M is such that $(P, M; C, D) = -1$, then M belongs to the polar of P.*

Proof of the Lemma. If the coordinates of C and D are $(1:0:0)$ and $(0:1:0)$, respectively, the equation of the conic is of the form $xy + z(kx + ly + mz) = 0$ (because the conic is not singular), k, l and m being constant, and the coordinates of P and M are of the form $(1:r:0)$ and $(1:s:0)$, respectively. By establishing the condition that P and M are polars, it turns out that $r + s = 0$, and with that, the condition of harmonicity holds. The converse theorem is obvious. □

The theorem is a consequence of this Lemma and that of the previous section, as is shown below.

Proof of the Theorem. In Figure 3.29 we denote the four points by A, B, C, and D, and we consider the lines AB, CD, AC, and BD. Let $P = AD \cap BC$, $Q = AC \cap BD$, and $R = AB \cap CD$. We shall prove that PQR is self-polar with respect to both conics.

By the lemma at the end of Section 3.8, $(P, M; B, C) = -1$, where $M = QR \cap BC$, and by the lemma we have just proven, M belongs to the polar of P. Analogously it is proven that $(R, N : A, B) = -1$, where $N = QR \cap BC$.

Hence N belongs to the polar of P, which implies that $MN = QR$, which is the polar of P. Similarly, it is proven that PQ is the polar of R, and then also PR is the polar of Q. Summarizing, the triangle PQR of the diagonal points of the quadrangle $ABCD$ is a self-polar triangle with respect to a conic containing the four points. □

Before closing this section, it is worthwhile mentioning that the condition imposed in the statement of the theorem arises naturally when the field of coefficients is $\mathbb{C}$, as established by a very important result about algebraic curves which we shall not prove (see [Fu]), but we shall explain, for it will allow us to prove that the degree of a polynomial equation is a projective invariant.

Theorem of Bezout. *Two projective curves in $P^2(\mathbb{C})$ without common components and with equations $F(x, y, z) = 0$, $G(x, y, z) = 0$, of degrees m and n, intersect at mn points.*

We know that if F and G define a projective locus, they must be homogeneous, and we also know that polynomials can be factored; therefore, the assumption that they do not have common components means that F and G do not have common factors. Under that condition, the theorem asserts that the number of common points of the curves defined by the equations $F(x, y, z) = 0$ and $G(x, y, z) = 0$, is mn.

Corollary. *The degree of a homogeneous polynomial with real coefficients, $F(x, y, z)$, is a projective invariant.*

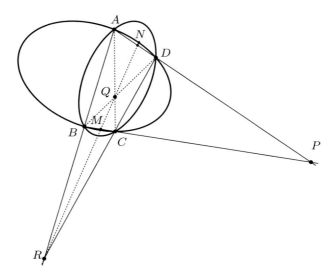

Figure 3.29: The diagonal triangle PQR of the quadrangle $ABCD$ is self-polar with respect to both conics.

Proof. We shall apply the Theorem of Bezout, taking $n = 1$, that is, when G represents a line. The number of points at which the curve defined by F intersects that line is the degree of F. In consequence any projectivity $T \in PGL(3, \mathbb{C})$ respects that number of intersections:

$$\#\{T(F) \cap T(G)\} = m,$$

for the common points are mapped into common points and there can not be new common points for the images, by the injectivity of T.

Since $PGL(3, \mathbb{R})$ is a subgroup of $PGL(3, \mathbb{C})$, the result is still valid when the polynomials F and G have real coefficients and $T \in PGL(3, \mathbb{R})$. Since the transform of a line is a line, the preservation of the number of intersections preserves the degree of F. □

Exercises

1. Prove that the polars of collinear points are concurrent.

2. Prove that a line through a point that lies inside of a conic (characterize them) cannot be tangent to the conic.

3. Prove that in the Euclidean plane $\mathbb{R}^2$ no line through the center of a hyperbola is tangent to the hyperbola, in spite of the center being exterior to the conic.

3.10 Elliptic geometry

To finish this chapter we study the **elliptic plane**. This means the projective plane $P^2(\mathbb{R})$ equipped with a specific rule for measuring lengths of curves (and areas of regions) in it.

To define this way of measuring in $P^2(\mathbb{R})$, we take into account that in S^2 we already have a way of measuring lengths and areas inherited from $\mathbb{R}^3$. Hence, we can define the **length of a curve** $\mathcal{C}$ of the elliptic plane as the length of any of the two curves in S^2 that are mapped into $\mathcal{C}$ under the standard projection

$$\Pi : S^2 \to P^2(\mathbb{R}).$$

As in Chapter 1, the curves must be smooth and parametrized, so that we can use the usual formula of calculus to compute its length:

$$l(\alpha) = \int_a^b ||\alpha'(t)||dt ,$$

where $\alpha : [a, b] \to S^2$ is a parametrization of the curve.

The problem of determining the curves on the sphere that minimize the path between two of its points, the **geodesics of the sphere**, was posed by navigation (let us forget about seas, islands, etc. to be surrounded) from remote times:

Which is the route from one point to another on Earth that minimizes the distance traveled?

To answer that question, we have recourse to an easy experiment which consists in fixing a tensed rubber band from A to B (see Figure 3.30).

If by keeping the extremes fixed we pull the rubber band and then we release it, the rubber band will always take the form of an arc of a **maximal circle**, that is, one of the circles that result from cutting the sphere with a plane through its center.

The Principle of the Minimal Effort asserts that this is the curve which minimizes the tension in the rubber band, because the length also does so.

In differential geometry [see DoC] it is said that a curve of a surface contained in $\mathbb{R}^3$ is a **geodesic** of that surface if at each one of its points, its acceleration vector is orthogonal to the plane tangent to the surface at that point.

That gives rise to a system of second-order differential equations determined by the geodesics when the initial conditions are fixed: one point through which the geodesic passes and its velocity. (The solution of that system of equations exists and is unique according to the Theorem of Existence and Uniqueness of Solutions of Differential Equations; see [L].)

The reason for asking the tangent component of the acceleration to be zero is inspired by Euclidean geometry: the lines of the Euclidean plane have this property. It is easy to prove that the curves of a surface in $\mathbb{R}^3$ with that property of zero acceleration, minimize the distance among sufficiently close points. Saying

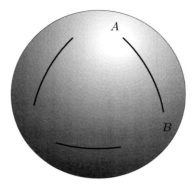

Figure 3.30: The geodesics of the sphere are maximal circles.

"sufficiently close" is justified if we notice that any two points in the sphere are in a maximal circle, which is divided in two arcs by these points, and if the points are not antipodal, then only one of these arcs provides the route of minimal length.

In other words, a geodesic in a surface S is a curve $\mathcal{C}$ in it, such that for each point $P \in \mathcal{C}$, there is a neighborhood U of P in S such that if P' is a point in $U \cap \mathcal{C}$, then $\mathcal{C}$ is the path in S of shortest length joining these two points.

In the Euclidean plane the straight lines are the geodesics, and they are used to form polygons; we established some of their properties in Chapter 1. Now we may ask ourselves what if we replace the euclidean plane and its straight lines by the sphere with its geodesic lines? Thus we arrive at **spherical geometry**.

Differently to what happens in the Euclidean plane, in the sphere it is false that two points determine a unique geodesic. For instance, through the poles of the Earth (or through any two antipodal points of the sphere) pass infinitely many different maximal circles, all the meridians.

However, that does not happen if instead of the sphere we consider the elliptic plane, for two projective lines intersect at only one point. This is an important reason for considering **elliptic geometry** instead of spherical geometry.

An **elliptic line** is therefore, by definition, a geodesic in the elliptic plane. These are the images in $P^2(\mathbb{R})$ of the circles of maximal length in the 2-sphere under the projection $\Pi : S^2 \to P^2(\mathbb{R})$.

Notice that the circles of maximal length in S^2 correspond to the intersections of the sphere with 2-planes in $\mathbb{R}^3$ passing through the center 0. Hence the projective lines actually become elliptic lines (or geodesics) when we endow $P^2(\mathbb{R})$ with the previous way of measuring lengths of curves. So these lines can be thought of as being formed by the representative elements of norm 1 of points of a projective line.

Since any two projective lines always intersect at one point, we can establish a first result:

3.10. Elliptic geometry

1. In the elliptic plane there are no parallel lines.

This fact is one of the possible denials of Playfair's Axiom, which is equivalent to Euclid's Postulate V; hence, elliptic geometry is a **non-Euclidean geometry.**

Another result in contrast to the Euclidean case is:

2. The elliptic lines have finite length (actually π), and the elliptic plane itself has finite area, 2π (half the area of S^2).

The first scholars who approached non-Euclidean geometries (see [Bo], [W] or [R-S]) arrived at the conclusion that by, substituting Postulate V for its denial $N1$: "a line does not admit parallels" (see Appendix 5.3), the lines should have finite length; this was enough for them to consider that they had arrived at an absurd result.

We leave to the reader the task of proving that in the elliptic plane Postulates I to IV hold, although Postulate II must be interpreted in the sense suggested by Bernhard Riemann: it is always possible to go forward on a geodesic, for as long as we wish.

Another very interesting result refers to the sum of the angles of an elliptic triangle, that is, a triangle whose sides are arcs of elliptic lines.

3. The sum of the angles of an elliptic triangle is greater than $180°$.

In Figure 3.31, a spherical triangle is shown to the left (and, necessarily, the antipodal triangle that is congruent), and the maximal circles which give rise to the elliptic lines are shown to the right.

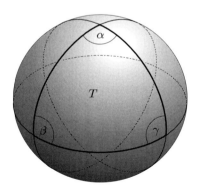 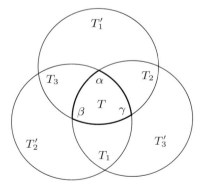

Figure 3.31: In an elliptic triangle the sum of its angles is greater than $180°$.

In both drawings of Figure 3.31 we denote by T the elliptic triangle, and by α, β and γ its interior angles (remember that the angle between two intersecting curves is the angle between the tangents to the curves at the point of intersection).

Let us prove the above statement, that the sum of the angles of T is more than $180°$. For this, let us call the part of S^2 between two meridians (maximal circles) forming an angle α a **spindle** of angle α of S^2. It is easy to see that such a spindle has area 2α (prove it using that the unit sphere has area 4π). Notice that if we cut this spindle by the meridian which is orthogonal to the other two, then the spindle splits in two congruent triangles of equal area α, whose inner angles are $\pi/2, \pi/2$ and α, so its sum is $\pi + \alpha > 180°$, proving the claim for these triangles.

In general, the triangle T is contained in the three spindles: one with angle α, another with angle β, and one more with angle γ.

In the spindle of angle α, the complement of T is denoted by T_1; in that of angle β, the complement is denoted by T_2, and in that of angle γ the complement is denoted by T_3. Therefore, if we use the same letter for the areas and for the triangles, we have

$$T + T_1 = 2\alpha\,; \quad T + T_2 = 2\beta\,; \quad T + T_3 = 2\gamma\,. \tag{3.25}$$

The sum of these three equations is

$$3T + T_1 + T_2 + T_3 = 2(\alpha + \beta + \gamma). \tag{3.26}$$

Observe now that the triangles in Figure 3.31 that have been denoted by T_1', T_2' and T_3' are congruent to T_1, T_2 and T_3, respectively, because they are antipodal (see the drawing to the left in Figure 3.31). That is why the following equality is valid. This formula expresses the area of the hemisphere illustrated by the bottom left disk in the right-hand side of Figure 3.31, :

$$2\pi = T + T_1 + T_2' + T_3 = T + T_1 + T_2 + T_3. \tag{3.27}$$

By substituting the value we have just obtained for $T + T_1 + T_2 + T_3$, it turns out

$$2T + 2\pi = 2(\alpha + \beta + \gamma), \tag{3.28}$$

and since $T > 0$ because it is the area of an elliptic triangle, we have proven that the sum of the angles of an elliptic triangle is greater than π. $\square$

Finally, let us notice that equation (3.28) establishes that

4. The sum of the angles of an elliptic triangle determines its area: If T has interior angles α, β and γ, then:

$$\text{Area}(T) = \alpha + \beta + \gamma - \pi\,.$$

This statement is not true in the Euclidean plane: the sum of the angles of a Euclidean triangle is constant, $180°$. On the other hand, we have similar triangles (with the same angles) whose areas are as small or as large as we wish. Actually, we can say more:

5. In the elliptic plane there are no similar triangles which are not congruent.

3.10. Elliptic geometry

To prove it, we use Figure 3.32, in which the drawing to the left illustrates the elliptic triangle of angles α, β and γ. The planes Π_1 and Π_2 through the origin, which determine the meridians forming the angle α, have normal vectors corresponding to two radii, r_1 and r_2, which form that same angle α, for the tangent lines with angle α and the radii r_1 and r_2 are perpendicular to the line $\Pi_1 \cap \Pi_2$.

We shall show how to determine a radius r_3, forming with radius r_1 an angle γ, and with radius r_2 an angle β.

The drawing to the right in Figure 3.32 illustrates the cone of revolution about r_2 whose generatrices form an angle β with r_2, and the cone of revolution about r_1 whose generatrices form an angle γ with r_1; remember that the angle between r_1 and r_2 is α.

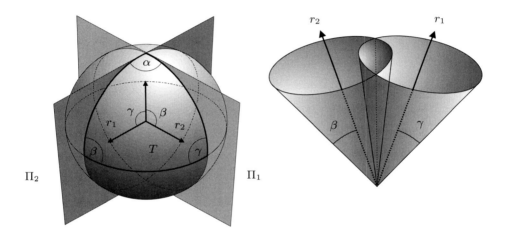

Figure 3.32: The angles of an elliptic triangle determine it (except for orientation).

The two cones intersect at two common generatrices, r_3 and r_3', symmetric with respect to the plane containing r_1 and r_2. Those generatrices determine the planes Π_3 and Π_3', which give rise to maximal circles forming, with those stemming from Π_1 and Π_2, two congruent triangles with opposite orientations, having angles α, β and γ, which concludes the proof. □

The results we have obtained so far are only a part of elliptic geometry, and the sphere itself has very important properties which we have not mentioned; to know several of them, the reader can see [H-C].

We only mention briefly the theme of the spherical tessellations, that is, determining all the possible ways of covering the sphere with regular and congruent spherical polygons.

The total of the possibilities results by projecting the Platonic solids into the sphere from its center, and by adding just the degenerated case of two hemispheres, each of which can be considered as a polygon of two sides (meridians) forming an angle of 180° (that is why we call it a degenerated case). The justification that there are no more possibilities is left as an exercise for the reader (see Exercise 4). We refer to sections II.1 and II.2 in [Se] for more on this topic.

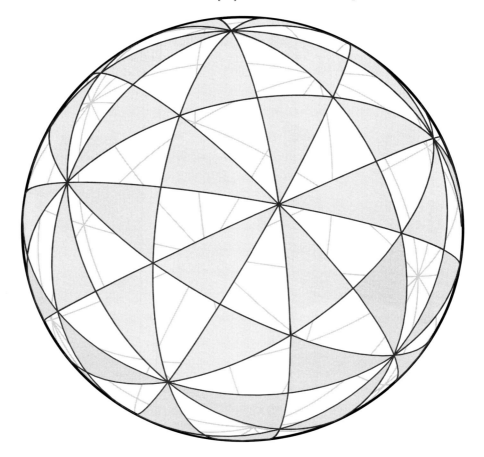

Figure 3.33: Tessellation of S^2 determined by a regular dodecahedron.

Exercises

1. Verify the validity of the four first Euclidean postulates in the elliptic plane.

3.10. Elliptic geometry

2. Determine the minimal upper bound for the sum of the angles of an elliptic triangle.

3. Determine the minimal upper bound for the length of an elliptic circle.

4. Prove that there are only five non-degenerated regular spherical tiles with which the sphere can be covered.

5. A Platonic solid is inscribed into a sphere S, but if we take a sphere with the same center and of radius smaller than that of S cutting each edge of the Platonic solid in two points, it is possible to use those points to obtain other tessellations of the smaller sphere, which although not formed by congruent spherical polygons only, have certain symmetries. Can you determine the subgroup of $O(3)$ that leaves each of them invariant?

4
Hyperbolic geometry

In Chapter 3 we saw that elliptic geometry is a non-Euclidean geometry, for any pair of elliptic lines intersect; that is, parallel lines do not exist in that geometry (denial $N1$ of Postulate V).

In this chapter, we study hyperbolic geometry, where denial $N2$ of Euclid's Postulate V holds: Given a line and a point which is not in the line, there exists more than one parallel to the line passing through the point. Again, we shall do it by using models, as in the cases of Euclidean and elliptic geometries.

The history of the discovery of the existence of this geometry, and of the creation of a good model for it, is very interesting (see [Ki], [R-S], [Y] or [W]); thus, the first section of this chapter presents the usual models of the hyperbolic plane. Surprisingly, these models are related to one another by means of bijective projections, and there exists a function that takes one into the other and allows using both of them without distinction.

To achieve the development of hyperbolic intuition it is very important to get familiar with the so-called conformal models (this concept will be explained below), by doing by oneself the respective drawings, although there is software which does it. Another source consists in analyzing M.C. Escher's engravings related to this theme; some of them are included in [Cox1] accompanied by an analysis, and many more appear in [Er] or [Es].

Hence, we invite the reader to observe how the use of complex coordinates gives rise to a very fruitful interrelation of algebraic, topological, metric, and analytic arguments, as well as to study, if he or she desires to go deeply into this fascinating theme of mathematics, the book [T] by one of the best contemporary geometers, William Thurston, whose works gave a great impulse to hyperbolic geometry in the 1980s.

4.1 Models of the hyperbolic plane

Once Janos Bolyai (1802–1860) and Nikolai Ivanovitch Lobachevsky (1793–1856) had proven the consistency of the system of axioms that results from substituting Postulate V in the Euclidean system by its denial $N2$ (Through a point not on a line there exists more than one parallel to the line), prominent mathematicians attempted to find a model of hyperbolic plane geometry among the surfaces contained in $\mathbb{R}^3$.

That meant, in particular, that the length of the curves on the surface, chosen as hyperbolic lines, the geodesics, should be infinite, when their length is measured as we do in calculus courses. The search was unsuccessful because, as David Hilbert (1862–1943) would prove later, it is impossible that there exists a surface of $\mathbb{R}^3$ with the required characteristics: that the way of measuring be the one induced by the usual metric of $\mathbb{R}^3$, that the geodesics have infinite length and that the Gaussian curvature, defined in Chapter 1, be constant and negative (see [DoC], Chapter 5).

A surface of $\mathbb{R}^3$ that has constant Gaussian curvature -1 is the **pseudosphere**, a surface of revolution generated by a **tractrix** (see Figure 4.1).

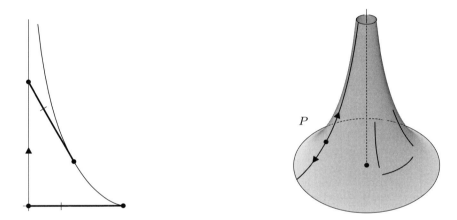

Figure 4.1: Geodesics of the pseudosphere.

An example of the tractrix is the curve drawn on the sand by a trolley toy pulled by a little boy, after he rotates 90° with respect to his initial direction (which we have marked with a vertical arrow in Figure 4.1); the projection of the small rope on the ground gives a tangent segment between the object and the perpendicular trajectory, which becomes an asymptote of the tractrix; since the small rope is taut, the segment has always the same length. The parametrization of the tractrix is proposed as Exercise 1, and the reader must verify that it satisfies this last condition which characterizes it.

In Figure 4.1 we have illustrated some geodesics of the pseudosphere; to obtain them, we can employ again a taut band, except that in this case there will be situations where the band must be placed inside.

The great flaw of this model, discovered by Eugenio Beltrami (1835–1900), is that it is **not complete**, that is, one can not walk as much as desired on one of the generatrices when going to the hole of the trumpet; that violates Postulate II.

Beltrami himself elaborated two other complete models. With the way of measuring that we shall introduce, lines will have infinite length. However, both

4.1. Models of the hyperbolic plane

models have the disadvantage of not being conformal; in other words, given two geodesics that intersect at a point, the angle we see, that formed by the tangent lines, is not the angle that corresponds to the way of measuring. However, both are useful for studying different problems (see [Ve]), and that is why we include them.

The **projective model** of the hyperbolic plane, also due to Beltrami, consists of a disk without the boundary. The points of the interior of the disk will be the **points of the hyperbolic plane**, and the points of the boundary of the disk are called **points at infinity**, since they play a role similar to the points at infinity of the affine plane.

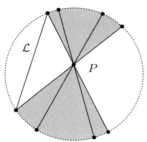

Figure 4.2: The projective model of the hyperbolic plane.

In Figure 4.2 we have drawn some chords; Beltrami called these chords **hyperbolic lines**, and it is clear that there are many of them containing the point P and that do not intersect the line $\mathcal{L}$: all the chords within the gray area. Thus, denial $N2$ of Postulate V holds (see Appendix 5.4).

Since the hyperbolic lines of this model are segments of Euclidean lines, actually we could think that we have done nothing new. The difference is the metric given for this model (see [Ve]), that it is not induced by the usual scalar product, and that has as a consequence, for example, that the hyperbolic lines have infinite length, that is, a point moving towards the boundary with constant speed will never reach it.

We shall very carefully construct the hyperbolic metric below by using another of the models.

The validity of Postulate I is immediate (see Appendix 5.4), for it is always possible to draw the segment corresponding to two given points of the hyperbolic plane. It is also true that we can produce that segment to obtain a complete hyperbolic line, that is, a cord that, as we shall see, turns out to have infinite length. Therefore, Postulate II will also be valid.

When we introduce the model in which we shall define the way of measuring lengths and angles, the reader will be able to verify that the lines with the same point at infinity are tangent in it (notice that each line has **two**, not one, points at infinity). That is not what we see in the projective model, for the chords that have a common point on the boundary form angles different from zero; therefore, in this model, one angle is that which we "read" when seeing it (since the only intuition we have developed is the Euclidean one), and another angle is the one that comes out from the corresponding formulas.

The reason for calling this model projective is that the group of transformations permitted in it is the subgroup of $PGL(3, \mathbb{R})$ obtained by fixing the conic $x^2 + y^2 - z^2 = 0$. To avoid misunderstandings caused by the term "projective," we shall call it the **Beltrami model**.

The second model is called the **model of the hyperboloid**, and it is obtained from the former by placing the disk on the plane $z = 1$ (we shall call this disk D_1), and projecting from the origin the points $P \in D_1$ into points P' of the upper sheet of the two-sheeted hyperboloid (see Figure 4.3)

$$-x^2 - y^2 + z^2 = 1.$$

By cutting the upper sheet of the hyperboloid with a plane through the origin, we obtain a branch of a hyperbola which is a hyperbolic line of this model.

The reader will agree that we are projecting (from the origin) the lines of the first model into the hyperboloid; this projection is bijective and, therefore, it is possible to find many lines $\mathcal{M}'$ in the model of the hyperboloid that pass through a point Q' exterior to a line $\mathcal{L}'$ without intersecting it.

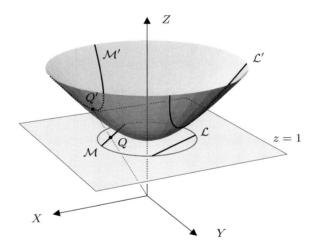

Figure 4.3: The model of the hyperboloid of the hyperbolic plane.

The other two models of the hyperbolic plane are due to Henri Poincaré (1854–1912), and they are the most convenient to develop the hyperbolic intuition because they are conformal.

4.1. Models of the hyperbolic plane

The **model of the Poincaré disk**, Δ, also consists of a disk without its boundary, but in this case the curves chosen as hyperbolic lines are the arcs of circle that intersect the boundary orthogonally.

In Figure 4.4 we have drawn several hyperbolic lines with the same point at infinity; those lines are called **hyperbolic parallels**, unlike **ultraparallels** such as $\mathcal{L}$ and $\mathcal{M}$, which are lines that do not intersect at points of the hyperbolic disk or at points at infinity.

Two hyperbolic parallel lines are tangent at a point at infinity, since all of them converge to the same point on the boundary, forming an angle of 90° with it. Notice how the pencil of parallels varies: to the right of the diameter that passes through the point at infinity P_∞, the circles are concave to the right, and those at the left are concave to the left.

The **model of the upper half plane**, H^+, has as points those of the upper half plane of the Cartesian plane: $P(x, y)$ with $y > 0$, and the points of the X axis play the role of points at infinity.

In this model, the hyperbolic lines are the semi-circles perpendicular to the X axis; among them we include those of infinite radius: the straight lines perpendicular to the X axis. Figure 4.5 is analogous to Figure 4.4.

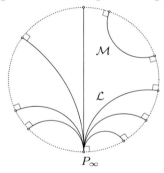

Figure 4.4: Model of the disk of Poincaré, Δ.

The model of the upper half plane has a physical origin that is explained as follows. When a spoon is submerged into a glass of water, by looking through the glass it seems that the spoon is bent; that is so because light does not follow the trajectory of a straight line when passing from one medium (air) to another (water), but it modifies its trajectory according to the following law.

Law of Refraction (W. Snell, 1621). *The product of the sine of the angle of incidence with the coefficient of density of the medium from which the light comes, is equal to the product of the sine of the angle of refraction times the coefficient of density of the medium into which it enters* (see Figure 4.6, where the angles are measured with respect to the perpendicular):

$$d_1 \sin \alpha = d_2 \sin \beta.$$

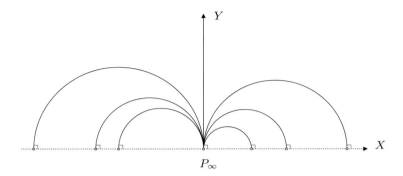

Figure 4.5: The model of the upper half plane of Poincaré, H^+.

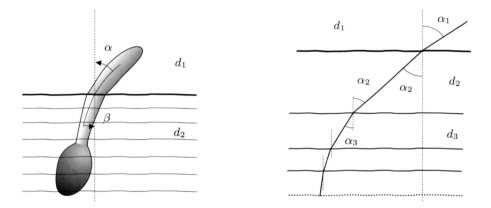

Figure 4.6: In a non-homogeneous medium, the trajectory of light does not follow a straight line.

At the right we have illustrated the trajectory of light when it successively traverses media whose density increases. The trajectory is not a line, for an essential property of light is that it travels so that the time elapsed is minimized, not the distance.

Notice that if the density increases exponentially (say 1 at height 1, 2 at height $1/2$, 2^2 at height $1/4$, and so on), light never reaches the border; that is why we call those **points at infinity** or **ideal points**. The union of all these points is called the ideal boundary. In the model of the disk, this ideal boundary is a circle that plays the role of the "horizon", and it is called "the visual sphere (or circle)".

We conclude this section by exhibiting bijections among the different models; since we already have a bijection between the Beltrami model and the model of the hyperboloid, it suffices to give a bijection between the Beltrami model and those of Poincaré.

4.1. Models of the hyperbolic plane

For that purpose, let us consider the unit sphere S^2 of $\mathbb{R}^3$, and let us place the projective model on the disk of the equator (see Figure 4.7 (a)).

Let us project the hyperbolic lines of the Beltrami model perpendicularly into the upper hemisphere of the sphere; each of the lines determines a plane perpendicular to the plane of the equator and, accordingly, it intersects the sphere in a circle *perpendicular to the disk of the equator*. This gives a "model" for hyperbolic geometry on the upper half sphere.

Now we use the stereographic projection of the sphere onto the plane of the equator from the point, $(0,0,1)$; the circles on the sphere which are perpendicular to the equator are transformed into circles of the XY plane and remain perpendicular to the equator, for the stereographic projection is conformal as the reader should have proven in Exercise 9, Section 3.1.

The composition of both projections, first XY on S^2 and then the stereographic projection of S^2 on XY, transforms the lines of the projective model into lines of the model of the Poincaré disk Δ.

In other words, we are doing the following: each line l in the Beltrami model meets the boundary at two "ideal" points, say a, b. Then the hyperbolic line in the Poincaré disk model that corresponds to l is the portion contained in the disk of the unique circle in the XY plane that meets the boundary circle perpendicularly at the points a, b.

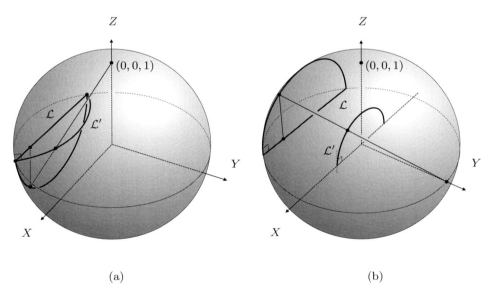

Figure 4.7: Relation between the different models of the hyperbolic plane.

To get the model of the half plane H^+, again we project the Beltrami model perpendicularly into the upper hemisphere, and then we project stereographically

S^2 on the XZ plane from the point $(0, 1, 0)$ (see Figure 4.7(b)).

The equator gives rise to the X axis, and the upper hemisphere is mapped into the upper half plane of the XZ plane. Because of the conformality of the stereographic projection, the semi-circles orthogonal to the equator are projected into semi-circles perpendicular to the X axis, giving rise to the model of the upper half plane of Poincaré.

Notice that we have established a bijection between the model of the disk, Δ, and that of the half plane, H^+; the reader should prove that it is conformal.

We suggest that the reader solve all the following exercises and pose some others; that is the only way to develop hyperbolic intuition.

To that respect, it is worthwhile noting that the model of the Poincaré disk has a very important advantage in contrast to the model of the half plane: it shows that all the points of the boundary have the same quality.

Instead, in the model of the half plane, the point that is missing in the X axis to make it a topological circle seems to play a special role; that is false. However, we shall follow the traditional notation and denote it with the symbol ∞.

Exercises

1. Verify that a parametrization of the tractrix is

$$\alpha(t) = \left(\sin t, \cos t + \log \tan(\frac{t}{2}) \right),$$

   where t measures the angle that the position vector of $\alpha(t)$ forms with the vertical axis (see Figure 4.1).

2. In the projective model, find two hyperbolic lines which contain the point $(0 : 0 : 1)$ and which do not intersect the line $2x - z = 0$.

3. Determine the hyperbolic line of the model of the hyperboloid, into which the line of the projective model that passes through the points $(1 : 0 : 1)$ and $(0 : 1 : 1)$ of D_1 is projected.

4. Prove that the bijection established between the two models of Poincaré is conformal.

5. In the model of the Poincaré disk , Δ, mark four points of the boundary so that they divide it into congruent arcs (in the Euclidean sense), and draw the hyperbolic lines that join two consecutive points. Now divide each arc into two equal parts, and draw the hyperbolic lines that join the new points with their adjacent points. Repeat the process while it is possible. The polygons one gets in this way are called **ideal**, since their vertices are points at infinity.

6. In the model of the Poincaré disk , Δ, mark eight points, $1, 2, \ldots, 8$, in the boundary so that they divide it into congruent arcs, and draw the lines of the model that join 1 with 3, 2 with 4, and so on. Observe that these hyperbolic lines determine a regular hyperbolic octagon. By taking into consideration that the models of Poincaré indeed are conformal, state the measurement of the angles at which any two of the lines

intersect, and the sum of the interior angles of this regular hyperbolic octagon.

7. Generalize the method used in the previous exercise to construct regular hyperbolic polygons with n sides for $n > 4$.

8. Design a process similar to the previous one to construct a regular hyperbolic triangle with no vertices at infinity.

9. Observe that in the second step of Exercise 5 you obtained an **ideal** octagon whose sum of interior angles is $0°$, while for the ordinary octagon of Exercise 6 the sum of the interior angles was greater than 2π. Can you give an argument to justify the existence of regular hyperbolic octagons whose sum of interior angles is exactly 2π?

10. In the model of the upper half plane, H^+, draw several triangles and in one of them verify that the sum of the interior angles is less than $180°$ (remember that angles are measured in the usual way).

11. In each of the models of Poincaré, draw a hyperbolic line $\mathcal{L}$ and, for each of its two points at infinity, draw several elements of the pencil of parallels to $\mathcal{L}$.

4.2 Transformations of the hyperbolic plane

Since we have several models of the hyperbolic plane, in each of them we must give transformations which are not just bijections between the points, but also bijections between the lines. Let us start with the Beltrami model; we leave the case of the hyperboloid as an exercise for the reader.

The Beltrami model is called projective precisely because its group of transformations is a subgroup of $PGL(3, \mathbb{R})$.

If we make the equation of the circle that is the boundary of the disk homogeneous, we obtain a projective conic $\mathcal{C}$,

$$x^2 + y^2 - z^2 = 0. \tag{4.1}$$

Consider the projective transformations $T \in PGL(3, \mathbb{R})$ that leave the conic $\mathcal{C}$ invariant (as a set); that is, $T(\mathcal{C}) = \mathcal{C}$. These form a subgroup $G_\mathcal{C}$ (prove it) and take the interior of the conic into itself because of what follows.

In Section 3.9, we saw that a conic separates $P^2(\mathbb{R})$ into two disjoint parts: a Möbius strip and a disk. The strip and the disk are not homeomorphic because in the disk every closed curve can be deformed into a point without leaving the disk, and in the strip there are curves, such as the central line, which do not have that property.

Thus, since any $T \in G_\mathcal{C}$ is the projectivization of a non-singular linear transformation, that makes it a homeomorphism; accordingly, T can not interchange the interior and the exterior of the conic. That is, a projective transformation T leaving the conic invariant necessarily leaves its interior invariant; hence, T maps points of the Beltrami model into points of that same set.

It is also true that hyperbolic lines are transformed into hyperbolic lines, since a plane through the origin of $\mathbb{R}^3$ is transformed by T into another plane through the origin, and if the plane cuts the interior of the conic, the transformed plane cuts it as well, according to the previous result.

If we take representatives $(x : y : 1)$, we obtain the disk D_1 of Figure 4.3, and the hyperbolic lines result from cutting D_1 with planes of $\mathbb{R}^3$ through the origin.

The reader is asked to verify that there always exists an element $T \in G_\mathcal{C}$ which transforms any point P of the hyperbolic plane into another point Q, both arbitrarily chosen, and the same happens with two hyperbolic lines.

Furthermore, we can fix a hyperbolic line $\mathcal{L}$ and a point L in it, and another hyperbolic line $\mathcal{M}$ and a point M in it, and it will always be possible to find a transformation $T \in G_\mathcal{C}$ that maps $\mathcal{L}$ into $\mathcal{M}$ and L into M. That is why it is said that the **group $G_\mathcal{C}$ is transitive at points and lines** (see Exercise 3).

For the model of the hyperboloid, the group of permitted transformations is the subgroup of $GL(3, \mathbb{R})$ that fixes the sheet of the hyperboloid with $z > 0$; the reader should characterize it.

Let us now see the models of Poincaré, where we make a more complete study of the group of permitted transformations and the associated invariants.

The advantage of having various models for the same geometry is that each model has certain advantages and gives different insights. So we warn the reader that we shall use different models without distinction, as it may be convenient.

As usual, we want to consider transformations that leave invariant the set we are dealing with, and also that they take hyperbolic lines into hyperbolic lines, that is, Euclidean circles (including straight lines) orthogonal to the boundary into other circles with the same property.

To begin with, notice that H^+ and Δ can be seen as subsets of $\mathbb{C}$, that is,

$$H^+ = \{z = x + iy \in \mathbb{C} \mid y > 0\}, \qquad \Delta = \{z \in \mathbb{C} \mid |z| < 1\}.$$

Moreover, observe that the transformation $Q \in PSL(2, \mathbb{C})$ (the group of Möbius transformations introduced at the end of section 3.5),

$$Q(z) = \frac{z - i}{-iz + 1}, \qquad (4.2)$$

maps the upper half plane H^+ into the disk Δ. Since we have Q explicitly, and its inverse, we are able to pass easily from either model to the other.

In Section 3.5 we proved that Möbius transformations are compositions of transformations that respect not only the angles, but also their orientation:

- homothetic map composed with rotation: $T(z) = cz$, with $c \in \mathbb{C}$;
- translations: $T(z) = z + b$;
- composition of the reflection with respect to the real axis with the inversion on the unit circle (see Appendix 5.6): $T(z) = 1/z$.

4.2. Transformations of the hyperbolic plane

Although that suffices to assure that a Möbius transformation necessarily maps a circle into another circle (admitting as circles those of infinite radius, the Euclidean lines), we shall prove this last assertion below with a simple computation.

Thus, the subgroup of $PSL(2, \mathbb{C})$ of maps that preserve the unit disk Δ, i.e., maps T such that $T(\Delta) = \Delta$, also take "lines" of the Poincaré disk model into lines of this model, and similarly for the subgroup of $PSL(2, \mathbb{C})$ of maps that preserve the upper half plane, for in both cases the corresponding lines are circles orthogonal to the boundary. These transformations preserve the angles and the orientations. They are also "isometries", as we shall see.

The transformations of each model that play the role of Euclidean reflections will be obtained by using a certain transformation, that somehow "reverses the plane" in question: it reverses the orientation of the angles and leaves the model invariant in each case.

In the case of Δ, the required transformation is the **conjugation**, which maps each complex number into its conjugate:

$$J : \mathbb{C} \to \mathbb{C} \quad \text{such that} \quad x + iy \mapsto x - iy.$$

Geometrically, the conjugate $\bar{z}$ of z results from reflecting z with respect to the real axis; that is why the conjugation reverses the orientation of angles.

The conjugation has also an algebraic interpretation: it is an **automorphism** of $\mathbb{C}$, that is, an isomorphism of the field $\mathbb{C}$ into itself, and it fixes each real number (see [B-ML]).

In the case of H^+, the transformation that reverses the orientation of angles that we shall add to obtain the inverse isometries will be the reflection with respect to the imaginary axis: $z \mapsto -\bar{z}$ (prove that z and $-\bar{z}$ are symmetric to one another with respect to the imaginary axis).

The conjugation also appears in the proof that the elements of $PSL(2, \mathbb{C})$ preserve "circles"; the reader should interpret the importance of this lemma in terms of the hyperbolic lines of the models.

Lemma. *Any Möbius transformation*

$$f(z) = \frac{az + b}{cz + d},$$

where $a, b, c, d \in \mathbb{C}$ and $ad - bc = 1$, maps circles into circles, considering as circles those of infinite radius, the Euclidean lines.

Proof. There are several ways for proving this result. One of them is by noticing, as before, that every Möbius transformation is a composition of simpler transformations, each of them obviously taking circles into circles. Here we use a different method: we use that we know how the equations of the straight lines and circles look to prove the lemma directly.

The equation of a "circle" $\mathcal{C}$ in the Cartesian plane is

$$Ax^2 + Ay^2 + Dx + Ey + F = 0. \tag{4.3}$$

Of course the coefficients are real, and it is clear that $A = 0$ when we are dealing with a Euclidean line.

By substituting in equation (4.3) the expressions for x and y as the real part x and the imaginary part y of z,
$$x = \frac{z + \bar{z}}{2}, \qquad y = \frac{z - \bar{z}}{2i}, \tag{4.4}$$
we obtain the equation of $\mathcal{C}$ in terms of z and $\bar{z}$ (make the computations):
$$Az\bar{z} + \frac{D - iE}{2}z + \frac{D + iE}{2}\bar{z} + F = 0.$$
This equation has the following properties: the coefficient of $z\bar{z}$ is a real number; the coefficient of z and that of $\bar{z}$ are conjugate; the independent term F is real. Since the coefficients are elements of $\mathbb{C}$, we shall write the equation of the circle $\mathcal{C}$ as
$$Az\bar{z} + Bz + \bar{B}\bar{z} + C = 0, \quad \text{with } A, C \in \mathbb{R}. \tag{4.5}$$
To obtain the equation of $f(\mathcal{C})$ we proceed as we have always done: since f has an inverse $f^{-1} \in PSL(2, \mathbb{C})$, if $f(z) = w$, then $z = f^{-1}(w)$, and we can substitute z and $\bar{z}$ in (4.5) by the expressions that we obtain for f^{-1}. For this, let us recall that the Möbius transformations can be handled in terms of matrices (the elements of $PSL(2, \mathbb{C})$):
$$\begin{pmatrix} a & b \\ c & d \end{pmatrix} \begin{pmatrix} z \\ 1 \end{pmatrix} = \begin{pmatrix} az + b \\ cz + d \end{pmatrix};$$
thus, as the determinant of the matrix is 1, its inverse (prove that this corresponds to f^{-1}) is
$$\begin{pmatrix} d & -b \\ -c & a \end{pmatrix}.$$
Hence,
$$z = f^{-1}(w) = \frac{dw - b}{-cw + a}.$$

The conjugate of a sum of complex numbers is the sum of the conjugates, and the analogous property holds for products; therefore, to obtain $\bar{z}$ it suffices to conjugate all the terms to the right, that is,
$$\bar{z} = \frac{\bar{d}\bar{w} - \bar{b}}{-\bar{c}\bar{w} + \bar{a}}.$$
The reader should verify that the expression that results from substituting z and $\bar{z}$ in (4.5),
$$A\left(\frac{dw - b}{-cw + a}\right)\left(\frac{\bar{d}\bar{w} - \bar{b}}{-\bar{c}\bar{w} + \bar{a}}\right) + B\left(\frac{dw - b}{-cw + a}\right) + \bar{B}\left(\frac{\bar{d}\bar{w} - \bar{b}}{-\bar{c}\bar{w} + \bar{a}}\right) + C = 0,$$
has the properties that characterize the equation of a circle when it is written in terms of w and $\bar{w}$. $\square$

4.2. Transformations of the hyperbolic plane

Corollary. *Four points z_1, z_2, z_3, z_4 in the extended complex plane $\widehat{\mathbb{C}}$ belong to a circle if and only if their cross ratio is real.*

Proof. The Fundamental Theorem of projective geometry is valid when the coordinates are complex numbers (review the arguments); therefore, there always exists a transformation $T \in PGL(2, \mathbb{C})$ which maps any three points of $P^1(\mathbb{C}) = \widehat{\mathbb{C}}$, z_1, z_2, z_3 into three other arbitrary points, w_1, w_2, w_3. If the latter points are on the real axis, the circle $\mathcal{C}$ through z_1, z_2 and z_3 is mapped into the real line and for any other point $z \in \mathcal{C}$, the cross ratio $(z_1, z_2; z_3, z)$ is real because $(w_1, w_2; w_3, T(z))$ is also. $\square$

To characterize the transformations of $PSL(2, \mathbb{C})$ that leave each model invariant, it suffices to require the oriented boundary of the considered set to remain invariant under the permitted transformations. In fact, by preserving the orientation of the angles, the region to the left of the boundary, when it is traversed as shown in Figure 4.8, is again mapped into the region to the left.

It is now clear that in either case, the transformations of $PSL(2, \mathbb{C})$ that leave each model invariant are a subgroup of $PSL(2, \mathbb{C})$. Let us determine what these subgroups are.

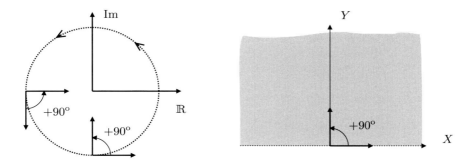

Figure 4.8: By respecting the boundary, a Möbius transformation maps the region to the left into itself.

We denote by G_{H+}^+ the subgroup of $PSL(2, \mathbb{C})$ that leaves the upper half plane invariant, and by G_Δ^+ the corresponding group for Δ. Let us determine G_{H+}^+ first.

If $f \in PSL(2, \mathbb{C})$ is such that $f(z) \in H^+$ for every $z \in H^+$, then one also has that f maps the boundary $\mathbb{R}$ into the boundary, by continuity.

It is immediate to verify that when all the coefficients of f are real numbers, we have $f(\mathbb{R}) \subset \mathbb{R}$, and since f preserves the orientation of $\mathbb{R}$, the upper half plane is mapped into the upper half plane.

We shall prove that the condition for the coefficients to be real is also necessary.

Let us then look for the subgroup of the transformations $f \in PSL(2, \mathbb{C})$ which map the real line into itself. By evaluating $f(z)$ at $z = 0$ we obtain

$$f(0) = \frac{b}{d} \in \mathbb{R};$$

and we can write $b = \lambda d$, with $\lambda \in \mathbb{R}$.

Since ∞ belongs to every line through the origin in $\widehat{\mathbb{C}}$, it must also hold that evaluating $f(z)$ at ∞ we obtain a real number, that is,

$$f(\infty) = \lim_{z \to \infty} \left(\frac{az + b}{cz + d} \right) = \frac{a}{c} \in \mathbb{R},$$

if $c \neq 0$ (what happens if $c = 0$?); accordingly, $a = \mu c$, with $\mu \in \mathbb{R}$. By evaluating $f(z)$ first at $z = 1$ and then at $z = -1$, the corresponding results are

$$f(1) = \frac{a + b}{c + d} \in \mathbb{R}, \quad f(-1) = \frac{-a + b}{-c + d} \in \mathbb{R}.$$

We leave as an exercise for the reader to prove that these relations imply that the coefficients a, b, c are real multiples of d; it is then possible to divide the numerator and the denominator of f(z) by d to obtain an expression where all the coefficients are real numbers. Moreover, if we further divide each coefficient by the square root of the determinant, we can have $a'd' - b'c' = 1$,

$$f(z) = \frac{a'z + b'}{c'z + d'},$$

and the transformation is unchanged. All this can be stated as the following theorem.

Theorem. *A Möbius transformation $f(z)$ leaves the upper half plane H^+ invariant if and only if it can be represented in the form*

$$f(z) = \frac{az + b}{cz + d},$$

with a, b, c, d real numbers such that $ad - bc = 1$. The group formed by these transformations is denoted as $PSL(2, \mathbb{R})$.

By adding the reflection with respect to the imaginary axis, we obtain the complete group of transformations of the upper half plane, G_{H^+}, whose associated geometry we shall study.

It can be proven that G_{H^+} contains any transformation that preserves angles (not necessarily their orientation) and leaves the upper half plane fixed, although we shall not do it in this book (see [Be]).

4.2. Transformations of the hyperbolic plane

Then, the **group of transformations that preserve angles and leave the upper half plane fixed**, G_{H^+}, consists of the transformations $f : \mathbb{C} \to \mathbb{C}$ of any of the following two forms:

$$f(z) = \frac{az+b}{cz+d} \quad \text{or} \quad f(z) = \frac{a(-\bar{z})+b}{c(-\bar{z})+d},$$

where $a, b, c, d \in \mathbb{R}$ and $ad - bc = 1$.

We can do something similar to determine the subgroup G_Δ that leaves the disk Δ invariant: characterize the elements of the subgroup of $PGL(2, \mathbb{C})$ that leave S^1 fixed, the boundary of the unit disk, and add the reflection with respect to the real axis.

However, it is easier to observe that it suffices to use the transformation Q in (4.2) and its inverse Q^{-1} to obtain, from the elements $f \in G_{H^+}$, the elements $h \in G_\Delta$:

$$h = QfQ^{-1} : \Delta \to \Delta. \tag{4.6}$$

It is an exercise for the reader to verify that QfQ^{-1} actually has as domain and co-domain Δ, and also to become convinced that each element $h \in G_\Delta$ can be expressed in this way, uniquely.

Then it is very easy to characterize the elements of G_Δ, since by multiplying the matrix of an element $f \in G_{H^+}$ by the matrix of Q and that of Q^{-1} following the order indicated by (4.6), the result is a matrix with complex entries where the elements of the diagonals are conjugate to one another.

We have then a theorem analogous to the previous one.

Theorem. *A Möbius transformation $f(z)$ leaves the unit disk Δ invariant if and only if it has the form*

$$f(z) = \frac{az+b}{\bar{b}+\bar{a}}, \quad \text{where } |a|^2 - |b|^2 = 1.$$

These transformations form a group.

Therefore, in the case of the hyperbolic disk, the group whose geometry we shall study, G_Δ, has elements of the form

$$h(z) = \frac{az+b}{\bar{b}z+\bar{a}} \quad \text{or} \quad h(z) = \frac{a\bar{z}+b}{\bar{b}\bar{z}+\bar{a}},$$

where $|a|^2 - |b|^2 = 1$.

Exercises

1. Prove that the set of transformations $T \in PGL(3, \mathbb{R})$ that leave the conic (4.1) fixed is a group G_C.

2. Find a transformation $T \in G_{\mathcal{C}}$ such that it maps the line $y = 0$ and the point $(5 : 0 : 1)$ into the line $x - y = 0$ and the point $(0 : 0 : 1)$. Solve this exercise in general.

3. Prove that if we arbitrarily choose a pair of lines $\mathcal{L}$ and $\mathcal{M}$ of the projective model, and fix a point on each, L and M respectively, it is possible to find $T \in G_{\mathcal{C}}$ such that it maps $\mathcal{L}$ and L into $\mathcal{M}$ and M. How many possible T's are there?

4. Determine the subgroup of $GL(3, \mathbb{R})$ that leaves the model of the hyperboloid invariant.

5. Prove that the map $z \mapsto \bar{z}$ from $\mathbb{C}$ into $\mathbb{C}$ is an isomorphism of the field $\mathbb{C}$ into itself that leaves $\mathbb{R}$ fixed point-wise.

6. Find the elements of $PSL(2, \mathbb{C})$ that correspond to each of the following transformations:
$f_1(z) = z + 1$; $f_2(z) = 2z$; $f_3(z) = (\cos\theta + i\sin\theta)z$; $f_4(z) = 1/z$.

7. Prove that the subgroup G_Δ that leaves the Poincaré disk invariant can be obtained from the subgroup G_{H+} by using the relation given by equation (4.6).

4.3 Steiner network

In this section we study essential properties of Möbius transformations which are very useful for studying the groups G_{H+} and G_Δ.

Let us start by observing that any Möbius transformation has two fixed points, α and β, which may be the same. That follows from the fact that when posing
$$\frac{az + b}{cz + d} = z,$$
we determine a second-degree equation,
$$cz^2 + (d - a)z - b = 0. \tag{4.7}$$

The Fundamental Theorem of Algebra asserts that both roots belong to $\mathbb{C}$, and to obtain them we apply the usual formula.

The geometric consequences of this algebraic fact are very interesting.

Let us first explore the case when $\alpha \neq \beta$; from the lemma in Section 4.2, it is immediate that any circle in the extended complex plane $\widehat{\mathbb{C}}$ through α and β is transformed under f into another circle passing through those points.

Let us call $\mathcal{F}_1$ the family of all the circles through α and β; according to what was stated above, $\mathcal{F}_1$ is invariant under f, that is, $f(\mathcal{F}_1) = \mathcal{F}_1$.

What is remarkable in this situation is that there is another family $\mathcal{F}_2$ which is invariant under f and whose elements intersect orthogonally with those of $\mathcal{F}_1$.

To visualize this second family, we assume that $\alpha = 0$ and $\beta = \infty$ (the transformation $h(z) = \frac{z-\alpha}{z-\beta}$ shows that there is no loss of generality) and we

4.3. Steiner network

identify these points with those corresponding to the north and south poles of the Riemann sphere (see Figure 4.9).

In the sphere, each element of the family $\mathcal{F}_1$ is formed by two meridians making an angle π; the elements of the family $\mathcal{F}_2$ are the parallels. It is clear that through each point P of the sphere, different from 0 and ∞, passes one and only one element of each family, and that the angle formed at P by these circles is a right angle.

To the right in Figure 4.9 there appears the result of stereographically projecting both families from $(0, -1, 0)$ on a plane through the north and south poles, denoted as N and S both in the sphere and in the plane.

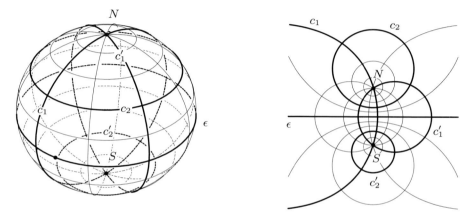

Figure 4.9: Any $f \in PSL(2, \mathbb{C})$ determines in $\widehat{\mathbb{C}}$ two invariant families of circles whose elements intersect orthogonally.

Since the stereographic projection is conformal, the elements of each family that intersect at a certain point of the plane different from N and S, do so orthogonally.

The reader familiar with this concept will suspect that the family $\mathcal{F}_2$ corresponds to the **circles of Apollonius** with limit points $\alpha = N$ and $\beta = S$, which we prove in Appendix 5.6. The set of these two families of circles forms a circular network on the plane called a **Steiner network**.

In Appendix 5.6 we prove other properties of the Steiner network related with inversion.

When $\alpha = \beta$ the result is a **degenerated Steiner network**; we invite the reader to draw it as a limit case of the previous one.

Now we want to study the effect of f on the families $\mathcal{F}_1$ and $\mathcal{F}_2$. This is simple when we take the model of the upper half plane; for the sake of simplicity we restrict ourselves to those transformations that preserve orientation, $f \in PSL(2, \mathbb{R})$. Thus, equation (4.7) has real coefficients and the study can be carried out using the fixed points of f, α and β, which are the roots of (4.7).

(a) The roots α and β are distinct real numbers; the transformation is called **hyperbolic** because it leaves two points at infinity fixed. Notice that the transformation $h(z) = \frac{z-\alpha}{z-\beta}$ maps $\alpha, \beta \in \mathbb{R}$ into 0 and ∞ respectively. This allows us to take f into a **standard form**, where the fixed points are 0 and ∞. In this case b and c must be both zero, so f is necessarily of the form

$$f(z) = kz, \text{ with } k > 0, \tag{4.8}$$

where $k = a/d$. This is called the standard (or canonical) form of a hyperbolic Möbius transformation. We ask the reader to verify that for each $k > 0$ there exists a matrix in $PSL(2, \mathbb{R})$ giving rise to such a transformation.

(b) The roots α and β coincide, and therefore they must be real. In this case the transformation is called **parabolic** because it leaves only one point at infinity fixed. The **standard form** of $f \in G_{H+}$ **parabolic** is obtained when the only fixed point is ∞. In this case the form of f is

$$f(z) = z + 1 \tag{4.9}$$

because, as before, ∞ being a fixed point implies that $c = 0$, and the fact that it is the only fixed point implies that the discriminant of equation (4.7) is zero; in consequence, $(a+d)^2 = 4$. Since additionally $ad = 1$, the previous equation becomes $(1/d) + d = \pm 2$, which implies $d = \pm 1$, and by projectivizing the resulting matrix we obtain $f(z) = z + k$, with $k \in \mathbb{R}$; to make the study, we take $k = 1$.

(c) The roots α and β are complex conjugate numbers; the transformation is called **elliptic** because it does not leave points at infinity fixed. In this case the **standard form** of $f \in G_{H+}$ is obtained by taking the fixed points to be i and $-i$. One gets:

$$f(z) = \frac{az + b}{-bz + a}. \tag{4.10}$$

Concerning the hyperbolic plane we conclude that the three possibilities are: (a) one has two fixed points in the ideal boundary, and the transformation is hyperbolic; (b) there is only one fixed point and it is in the ideal boundary, and the transformation is parabolic; or (c) there is one fixed point in the interior of the hyperbolic plane, and the transformation is elliptic.

We can also obtain interesting conclusions about the way these three types of transformations move the points of the hyperbolic plane (in either model), and their geometry around each fixed point

In case (a), the transformation $f \in PSL(2, \mathbb{R})$ leaves two points of the boundary fixed. Hence the hyperbolic line $\mathcal{L}$ through them also remain fixed (as a set); the other elements of the family $\mathcal{F}_1$ also remains fixed as sets, though they are not hyperbolic lines anymore; the points in them, other than the two fixed points, move along these curves, getting further and further away from one of the fixed points, and getting closer and closer to the other fixed point. That is why one of the fixed points is said to be an **attractor**, meaning that $f(P)$ is closer than P to the fixed point, while the other fixed point is a **repelling point**, for the points recede from it under f (see Appendix 5.6).

The elements of the family $\mathcal{F}_2$ are all hyperbolic lines; they are transformed one into another by f; by continuity, if one of these lines (perpendicular to the previous lines) is moved in one direction (toward the right or the left), all the other lines move in that same direction (see Figure 4.10 (a)).

In case (b), f leaves only one point of the boundary fixed ($\alpha = \beta$). The family $\mathcal{F}_1$ consists of Euclidean circles tangent to one another (and tangent to the real axis) at the fixed point; they are not hyperbolic lines and in hyperbolic geometry they are called **horocycles**. As we will see, they turn out to be curves equidistant to one another under the hyperbolic metric, that we define below, and each is invariant (as a set) under f. The points in the plane move along these lines. The latter is obvious if we consider the standard form $f(z) = z + 1$, whose fixed point is ∞ and where the horocycles are Euclidean lines parallel to the real axis. The family $\mathcal{F}_2$ does consist of hyperbolic lines, all of them tangent at α, and therefore they form a pencil of parallels. As in the previous case, if f moves them in one direction, for example to the left, the others will also move in that same direction (see Figure 4.10 (b)).

Finally, in case (c), since $\mathbb{R}$ is mapped into itself, the other elements of the family $\mathcal{F}_2$ (which are not hyperbolic lines) also remain invariant (as a set), while those of the family $\mathcal{F}_1$ (which are hyperbolic lines) are transformed one into another, giving the impression of turning around the fixed point (see Figure 4.10 (c)). Observe that the matrix of the standard form is a matrix of a rotation.

Here it is worthwhile emphasizing an algebraic fact.

Remark. The transformations $f \in PSL(2, \mathbb{R})$ can be classified according to the square of their trace, $a+d$, because the sign of the discriminant of (4.7) determines the number of real roots. However, if we compute the discriminant and recall that $ad - bc = 1$, it is easy to prove the following (see Exercise 2):

- $f \in PSL(2, \mathbb{R})$ is hyperbolic if and only if $(a + d)^2 > 4$;
- $f \in PSL(2, \mathbb{R})$ is parabolic if and only if $(a + d)^2 = 4$:
- $f \in PSL(2, \mathbb{R})$ is elliptic if and only if $(a + d)^2 < 4$.

The reader should make the drawings analogous to those of Figure 4.10 for the model Δ. For this topic, the classic reference is [Be].

Exercises

1. Analyze how the Steiner network is transformed under Q^{-1}, and make the drawings corresponding to Figure 4.10 in the case of Δ.

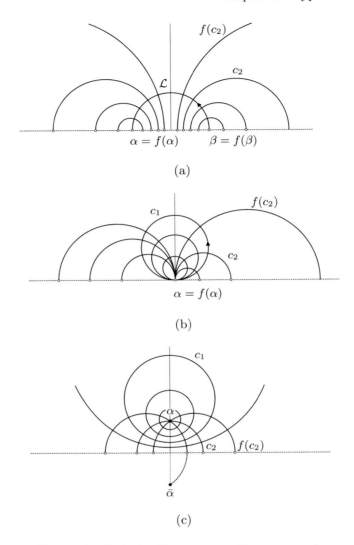

Figure 4.10: Each $f \in G_{H^+}$ defines a Steiner network.

2. Prove the assertion made in the final Remark.

4.4 The hyperbolic metric

We have already seen that in order to find models for hyperbolic plane geometry, we take certain surfaces, either in the plane or in $\mathbb{R}^3$, and then modify the concept

4.4. The hyperbolic metric

of "straight lines" (or geodesics, properly speaking). This corresponds to changing the way of measuring lengths of curves and leads to a different metric on these surfaces, the hyperbolic metric. The geodesics, or simply hyperbolic lines, are the curves that measure the hyperbolic distance between any two of its points; these are the curves of shortest length between any two points in the hyperbolic plane.

A consequence of Hilbert's result, mentioned in Section 4.1, is that to obtain the hyperbolic plane as a subset of some Euclidean space $\mathbb{R}^n$, with the metric induced by the usual scalar product, it is necessary that $n \geq 4$. Exercise 6 of this section will help to convince the reader that the hyperbolic plane "does not fit" in $\mathbb{R}^3$ if we want its metric to be induced from that in the ambient space.

The physical motivation for the model of the upper half plane posed that when the medium becomes denser as we approach the boundary, light, whose speed is constant, must modify its trajectory to keep on traveling at the same speed. If the density tends to infinite as we approach the boundary, light will never reach the border. This is what happens in the hyperbolic plane, and that is why the length of the geodesics is infinite.

A natural way to make the ambient space denser as we approach the boundary consists in modifying the Euclidean scalar product (which is basic to measure distances and angles) so that the norms are increased as we approach the boundary. The effect is that as we get close to the boundary, a step that may look small to the "Euclidean eyes", becomes large when we see it with "hyperbolic eyes".

Definition. The **hyperbolic scalar product** of two vectors in the upper half plane, $\bar{u}$ and $\bar{v}$, with the initial point based at $z = x + iy$, is given by

$$\bar{u} \cdot_H \bar{v} = \frac{\bar{u} \cdot_E \bar{v}}{y^2},$$

where $\cdot_E$ denotes the usual scalar product of the Euclidean plane (see Figure 4.11).

Notice that this hyperbolic scalar product inherits from the usual scalar product the properties needed for a scalar product:

- Its value is a real number for any two vectors $\bar{u}$ and $\bar{v}$ with initial point based at the same point; it is not negative when $\bar{u} = \bar{v}$ and, in this case, becomes zero only if $\bar{u} = \bar{0}$.

- It is distributive on addition of vectors, and it takes out scalars of each factor (some form of associative law); that is, it is bilinear.

- Its value does not depend on the order of the factors (obviously a commutative law); that is, it is symmetric.

Once the scalar product $\cdot_H$ is defined, we can define the **hyperbolic norm** of a vector $\bar{v}$ with initial point at certain point $P_0(x_0, y_0)$ of the upper half plane:

$$||\bar{v}||_H = (\bar{v} \cdot_H \bar{v})^{1/2}.$$

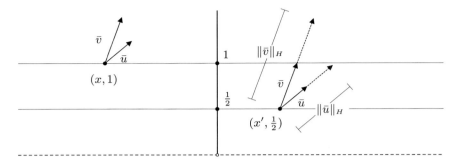

Figure 4.11: The hyperbolic scalar product in H^+ depends on the height where the initial points of the vectors fall.

With that it is possible to measure the hyperbolic length of a curve in H^+ analogously to the usual way: if $\alpha : [a, b] \to H^+$ is a differentiable curve, its **hyperbolic length** is

$$l_H(\alpha) = \int_a^b ||\alpha'(t)||_H dt.$$

Notice that the way of computing the norm varies with the point $\alpha(t)$ which is the initial point of the tangent vector; however, if the curve is smooth, the variation is continuous and that gives the integral meaning.

Given two points $P_1, P_2 \in H^+$, **the hyperbolic distance** of P_1 to P_2 is the infimum of the lengths of the curves in H^+ joining these two points.

Hence, we have a new way of measuring in the upper half plane, called a **hyperbolic metric**. The metrics obtained from scalar products that vary smoothly with the initial point of the vectors are called **Riemannian metrics**, and they play a central role in differential geometry.

To get an insight into what this metric does, imagine there is a person A that belongs to the hyperbolic plane and moves around; of course its "hyperbolic size" is constant. However, if we belong to the Euclidean world and we look at him from the outside, we see him in his true size only when he is standing near the hight $y = 1$ in H^+ (since at these points the hyperbolic and the Euclidean metrics coincide). As he walks down towards the real axis, he looks smaller and smaller to us, becoming just like an ant as he approaches the border. Conversely, he looks like a giant to us as he moves up, getting further away from the x-axis.

Our main purposes for giving this definition of scalar product is to define a metric for which the hyperbolic lines, as defined before, minimize the distance between two of its points, and they have infinite length.

Let us now prove the second statement in the special case of the positive Y semiaxis; then we will prove that by transforming this line into any other hyperbolic line using an element in $f \in G_{H^+}$, the hyperbolic norm of the tangent vectors does not change. That means that f is an isometry, and that any line has

4.4. The hyperbolic metric

infinite length.

The hyperbolic line corresponding to the positive Y semiaxis has a very simple parametrization:
$$(x(t), y(t)) = (0, t),$$
where $t \in (0, \infty)$. To prove that it has infinite length is enough to prove that the length of the half line corresponding to $(0, 1]$ is infinite. Recall that the way of computing the length of a curve is analogous to the usual one, except that the norm of the tangent vector is the hyperbolic norm:

$$\begin{aligned}
\int_0^1 \|(x'(t), y'(t))\|_H \, dt &= \lim_{a \to 0} \int_a^1 \|(0, 1)\|_H \, dt \\
&= \lim_{a \to 0} \int_a^1 \frac{1}{t} \, dt \\
&= \lim_{a \to 0} \ln(t)|_a^1 = \lim_{a \to 0} (0 - \ln(a)) = +\infty. \quad (4.11)
\end{aligned}$$

Accordingly, the length of this half hyperbolic line (which is a segment of Euclidean length equal to 1) is infinite.

Let us map the positive part of the Y axis into any hyperbolic line; it will suffice to consider the case of the semi-circle with center at the origin and of radius 1, additionally requiring i to be transformed into itself and the orientation to be preserved.

Thus, $f \in G_{H+}$ must satisfy
$$f(0) = -1; \quad f(\infty) = 1; \quad f(i) = i. \quad (4.12)$$

The elements of G_{H+} that preserve the orientation have the form
$$f(z) = \frac{az + b}{cz + d},$$
with $a, b, c, d \in \mathbb{R}$, and by (4.9) we have that
$$\frac{b}{d} = -1; \quad \frac{a}{c} = 1; \quad \frac{ai + b}{ci + d} = i.$$

Hence,
$$b = -d; \quad a = c; \quad ai + b = di - c, \text{ which implies } a = d, \ b = -c.$$

Therefore, the expression for $f(z)$ is
$$f(z) = \frac{dz - d}{dz + d} = \frac{z - 1}{z + 1}.$$

The points of the positive part of the Y axis have the form $z = it$; thus, by applying f to them we obtain
$$f(it) = \frac{it - 1}{it + 1} = \frac{(it - 1)(-it + 1)}{(it + 1)(-it + 1)} = \frac{-1 + t^2 + 2ti}{1 + t^2}.$$

That is, the points $f(it)$ on the semi-circle have the parametrization

$$(x(t), y(t)) = \left(\frac{-1+t^2}{1+t^2}, \frac{2t}{1+t^2}\right),$$

and the tangent vector is, after simplifying,

$$(x'(t), y'(t)) = \left(\frac{4t}{(1+t^2)^2}, \frac{2(1-t^2)}{(1+t^2)^2}\right).$$

If the reader computes the hyperbolic norm of $(x'(t), y'(t))$, he or she will find out that it is $1/t$, the same as that for the vector tangent to the positive Y semiaxis at the point it. We also leave to the reader the proof of the validity of the result for any other hyperbolic line, that is, when the circle has its center at any point of the X axis and its radius is arbitrary.

With that, the reader will have proven that the length of any segment remains invariant under any $f \in G_{H+}$, and we can state this result as follows.

Theorem 1. *The elements of G_{H+} are isometries of H^+ with respect to the hyperbolic metric.*

Also from the first computation another fundamental result is immediate:

Theorem 2. *Hyperbolic lines are curves that minimize the hyperbolic distance between two of its points. That is, they are geodesics.*

Proof. By Theorem 1, it suffices to consider the case of the hyperbolic line given by the imaginary semiaxis; any curve $(x(t), y(t))$ joining the points Ai and Bi of the imaginary axis has hyperbolic length greater than or equal to that corresponding to the segment from Ai to Bi, which is $L(B/A)$. To verify it, we consider the case of a curve without self-intersections with $t \in [a, b]$, and we recall that $y(t) > 0$:

$$\begin{aligned}
\int_a^b \|(x'(t), y'(t))\|_H \, dt &\geq \int_a^b \|(0, y'(t))\|_H \, dt \\
&= \int_a^b \left|\frac{y'(t)}{y(t)}\right| dt \\
&= \ln(y(t))|_a^b = \ln(B) - \ln(A) = \ln(B/A).
\end{aligned}$$

$\square$

Now the results come one after another as in a cascade, as the following.

Proposition 1. *Two horocycles with the same point at infinity are equidistant curves. That is, if from one point of one of the horocycles we draw a perpendicular to the other, the length of the segment of perpendicular is independent of the point.*

4.4. The hyperbolic metric   173

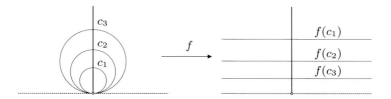

Figure 4.12: The horocycles with the point at infinity in common are equidistant curves.

Proof. A transformation $f \in G_{H+}$ that maps the fixed point to ∞ transforms any of the horocycles into a circle of infinite radius, that is, a Euclidean line. In addition, since all horocycles intersect orthogonally the hyperbolic line perpendicular to the X axis at the point of tangency, by transforming them into Euclidean lines these turn out to be horizontal (see Figure 4.12). Thus, the hyperbolic distance between two of them is constant, as was stated. □

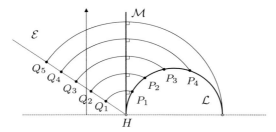

Figure 4.13: $\mathcal{L}$ and $\mathcal{M}$ are not equidistant curves, but $\mathcal{M}$ and $\mathcal{E}$ are.

Notice that two parallel hyperbolic lines, that is, with a point at infinity in common, are **not** equidistant curves, for the hyperbolic distance between them (the length of the segment of the perpendicular from $P \in \mathcal{L}$ to $\mathcal{M}$) increases as the point P recedes from the point at infinity.

In Figure 4.13 we illustrate the case when one of the hyperbolic lines is a parallel $\mathcal{M}$ to the imaginary axis: the distance from $\mathcal{L}$ to the line $\mathcal{M}$ increases as much as desired as P approaches the other point at infinity of $\mathcal{L}$; instead, any Euclidean half line $\mathcal{E}$ through H indeed is a curve equidistant from $\mathcal{M}$. The reader can prove it by noticing that all the arcs from Qi to $\mathcal{M}$ are homothetic.

We have not yet analyzed what a hyperbolic circle looks like; the concept makes sense since, by fixing a point C, in any direction from it we can draw a hyperbolic line and measure on it any distance r fixed in advance (see Figure 4.14).

We leave as an exercise for the reader to prove the following proposition. Figure 4.14 suggest how to proceed.

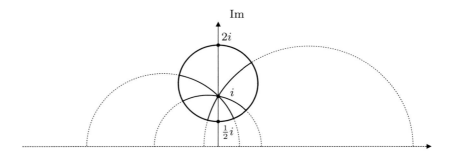

Figure 4.14: Hyperbolic circles are Euclidean circles, but the centers are different.

Proposition 2. *Hyperbolic circles are Euclidean circles, although the hyperbolic center is shifted toward the boundary.*

Now we can prove a theorem that perhaps the reader already expected.

Theorem 3. *The only isometries of H^+ into itself that preserve orientation are the elements of $PSL(2,\mathbb{R})$.*

Proof. Theorem 1, which was proven in this section, implies that the elements of $PSL(2,\mathbb{R})$ are isometries of H^+. Therefore, to prove Theorem 3 we only need to prove the implication in the other direction, that is, that these are all the isometries that preserve orientation.

As in the Euclidean and the elliptic cases, an isometry is a bijection $T : H^+ \to H^+$ which preserves the distance between points. Therefore, if a curve $\mathcal{C}$ is a geodesic, $T(\mathcal{C})$ is a geodesic as well. The argument used in the Euclidean case to verify that two congruent triangles determine a unique isometry T is also valid in this case, for if P, Q, R and P', Q', R', are the vertices of the two congruent triangles, any $S \in H^+$ determines with P and Q a triangle.

An isometry T that maps ΔPQR into $\Delta P'Q'R'$, also maps ΔPQS into $\Delta P'Q'S'$, and for S' there are only two possibilities, which are the points of intersection of the hyperbolic circles with center at each of the vertices P' and Q' and radii $d_H(P,S)$ and $d_H(Q,S)$, respectively. The choice will depend on whether $DPQR$ into $DP'Q'R'$ are equally oriented or not. If T respects the orientation, it necessarily coincides with the only element $f \in PSL(2,\mathbb{R})$ such that $(P,Q;R,S) = (P',Q';R',f(S))$. Otherwise, we must compose f with the reflection on the imaginary axis to obtain T. □

This section concludes by stating what is the metric in the case of the disk and expressing the hyperbolic distance in terms of the cross ratio.

We had stated that by giving a metric for one of the models, we would be able to define the metric for another model using a bijection between both models.

4.4. The hyperbolic metric

By using the bijection established by (4.2),

$$Q(z) = \frac{z-i}{-iz+1};$$

it is possible to prove that the scalar product for the model of the disk is given by the formula

$$\bar{u} \cdot_\Delta \bar{v} = \frac{4\bar{u} \cdot_E \bar{v}}{(1-r^2)^2}, \tag{4.13}$$

where r is the distance from the point of origin of the vectors to the center of the disk (see Figure 4.15).

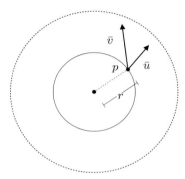

Figure 4.15: How to measure in the model of the disk of Poincaré.

The hyperbolic distance from the center of Δ to any point of the boundary is infinite, for we already proved that the distance from i (which is mapped into 0 by Q) to any point of the edge of the half plane is infinite.

Moreover, since 1 and -1 remain fixed, the hyperbolic line of H^+ that contains i, 1 and -1 is transformed into the hyperbolic line of Δ through 0, 1 and -1: the horizontal diameter. The remaining hyperbolic lines through i in H^+, give rise to the other diameters, all of them hyperbolic lines of Δ.

Let us now express the distance between two points z and w in terms of the cross ratio $(z^*, w^*; z, w)$, where z^* and w^* are the points at infinity of the unique hyperbolic line $\mathcal{L}$ determined by z and w (see Figure 4.16).

Let us consider the transformation $T \in PSL(2, \mathbb{C})$ that maps $\mathcal{L}$ into the imaginary axis so that z^* is transformed into 0; the other point at infinity is transformed into ∞ and the points z and w have images of the form iy_1 and iy_2.

The reader will have no trouble in justifying each of the following equalities:

$$\begin{aligned} d_H(T(z), T(w)) &= \ln\left(\frac{T(z)}{T(w)}\right) \\ &= \ln(0, \infty; iy_1, iy_2) \\ &= \ln(z^*, w^*; z, w). \end{aligned} \tag{4.14}$$

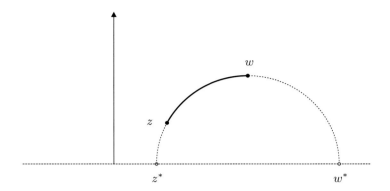

Figure 4.16: The distance from z to w is given by $(z^*, w^*; z, w)$.

Since the transformation $Q : H^+ \to \Delta$ is a projectivity and, therefore, leaves the cross ratio invariant, the previous formula is also valid for points in Δ.

Exercises

1. Prove that the Euclidean half line $\mathcal{E}$ of Figure 4.13 is a curve equidistant from the hyperbolic line $\mathcal{M}$.

2. Prove that two Euclidean rays with origin at the same point at infinity of H^+ are equidistant curves with the hyperbolic metric.

3. Is the Euclidean translation $z \mapsto z + b$, being $b \in \mathbb{R}$, a translation on the hyperbolic plane? Justify your answer.

4. Prove Proposition 2. (Suggestion: Locate two points A and B of the imaginary axis at the same hyperbolic distance r from i, and determine the Euclidean circle $\mathcal{C}$ of diameter AB. Given any hyperbolic line $\mathcal{L}$ through i, compute the hyperbolic length of the segment from i to $\mathcal{L}$.)

5. Justify formula (4.13).

6. Prove that there exist regular hyperbolic polygons of n sides with all their angles being right angles if $n > 4$.

7. Construct several regular Euclidean hexagons, all of them congruent, and by cutting along one radius, add to each a triangle congruent to the six triangles that form the hexagon. Now join the hexagons to another one along one of its sides; if the number of hexagons modified is large, it is not easy to handle the object, in spite of having negative curvature only at the centers of the "hexagons". This illustrates the impossibility of the hyperbolic plane fitting in $\mathbb{R}^3$ with the induced metric.

8. Prove, from (4.14), that d_H satisfies the triangle inequality.

4.5 First results in hyperbolic geometry

In this section we study several results obtained by the early scholars of hyperbolic geometry. This give us a taste of richness of this topic, which can be studied with many techniques. Some of the results we describe here contrast with what happens in the other geometries we have studied.

We do not present the original proofs, which can be found in [Bo] and [W]; a brief account of their historical development can be found in [R-S]. We use both of the Poincaré models without distinction.

1°. (Girolamo Saccheri (1667–1733)). *The sum of the angles of a hyperbolic triangle is less than* $180°$.

Proof. We use an isometry to map the vertex A to the center of the disk, and thus two sides are radii, while the other two vertices determine the side BC, which belongs to an arc perpendicular to the edge of the disk (see Figure 4.17). The hyperbolic angles β and γ are smaller than the respective Euclidean angles β' and γ'. Hence,
$$180° = \alpha + \beta' + \gamma' > \alpha + \beta + \gamma \,.$$ □

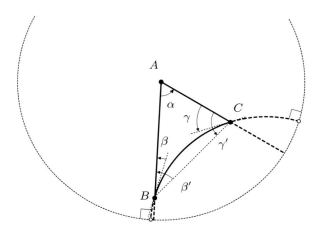

Figure 4.17: In a hyperbolic triangle, $\alpha + \beta + \gamma < 180°$.

It is interesting to compare the following result with the equivalent statement for triangles in spherical (or elliptic) geometry, explained in Chapter 3 and recalled below.

2°. (Johann Heinrich Lambert (1728–1777)). *The area of a hyperbolic triangle is equal to π minus the sum of the interior angles of the triangle.*

Actually, the original statement by Lambert refers to a hyperbolic polygon: "The difference of the sum of the interior angles of a Euclidean polygon minus the sum of the interior angles of a hyperbolic polygon with the same vertices is proportional to the (hyperbolic) area of the polygon".

Proof. In the case of an elliptic triangle, the computation of its area gave as result

$$\text{area of the triangle} = \alpha + \beta + \gamma - \pi;$$

in this case, the computation of the area of a triangle gives us the same type of information, although the arguments we use are entirely different.

Let us start by clarifying how to measure hyperbolic areas. It suffices to consider that, if the way of measuring the hyperbolic length of a vector tangent at the point $z = x + iy \in H^+$ varies with respect to the Euclidean length inversely to the imaginary part y, then the hyperbolic element of area varies with respect to that of the Euclidean inversely to y^2.

Since any $f \in G_{H^+}$ is a hyperbolic isometry, the area does not vary if we map the side AB of the triangle so that it coincides with the line corresponding to the semi-circle $|z| = 1$ (see Figure 4.18).

Let us make the computation in the case of the ideal triangle $AB\infty$ (notice that the height is infinite), for any triangle can be obtained as the difference of two triangles with the same ideal vertex (Exercise 1).

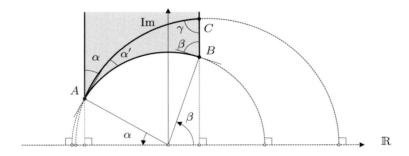

Figure 4.18: Area $(\triangle ABC) = \pi - (\alpha + \beta + \gamma)$.

It is clear that x varies from $\cos(\pi - \alpha)$ to $\cos\beta$, and that given x, the imaginary part y varies from $\sqrt{(1-x^2)}$ to ∞. Therefore, the area of the hyperbolic triangle $\triangle AB\infty$ is

$$\text{area } \triangle AB\infty = \int_{\cos(\pi-\alpha)}^{\cos\beta} \left(\int_{\sqrt{(1-x^2)}}^{\infty} \frac{dy}{y^2} \right) dx = \pi - (\alpha + \beta).$$

The reader can apply this result to the triangles $\triangle BC\infty$ and $\triangle CA\infty$ to obtain the statement that

$$\text{area } (\triangle ABC) = 180° - (\alpha + \beta + \gamma). \qquad \square$$

4.5. First results in hyperbolic geometry

Notice that, according to the previous computation, in hyperbolic geometry it is possible to have a triangle of infinite height and finite area.

In the case of a hyperbolic quadrilateral $\mathcal{P}$, with angles α, β, γ, δ, Lambert's result states that
$$\text{area } \mathcal{P} = 2\pi - (\alpha + \beta + \gamma + \delta).$$

The following result, due to Gauss, characterizes the circles in a way that is valid for hyperbolic circles and for Euclidean circles as well.

For that, Gauss defined when a point A of the hyperbolic plane is a point corresponding to a fixed point D **with respect to a pencil of hyperbolic lines with vertex** O. The condition is that the hyperbolic triangle OAD is isosceles, which Gauss expressed by saying that OA and OD must form equal angles on the same side of AD (see Figure 4.19).

3°. (Carl Friedrich Gauss (1777–1855)). *A circle is the locus of the points in the lines of a pencil corresponding to a fixed point D.*

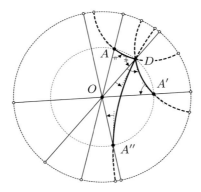

Figure 4.19: Points corresponding to D in straight lines through O.

Proof. In this case, we take the model of the disk and, by means of an isometry, we locate the vertex of the pencil at the center O of the disk. For an arbitrary fixed point $D \in \Delta$, the point corresponding to D in one of the lines of the pencil is the point A such that OAD is an isosceles triangle; for that reason not just the angles have the same amplitude, but also all the segments OA have the same length, $d_H(O, D)$ (see Figure 4.19). □

The following result should not surprise us after the remark made in the proof of **1°**, but it has the merit of defining a number associated to hyperbolic geometry.

4°. (Ferdinand Karl Schweikart (1780–1859)). *The height of a rectangle and isosceles triangle increases as the sides increase but without exceeding a certain length, called* **the constant**.

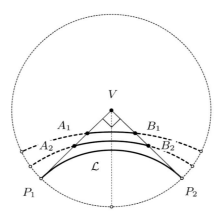

Figure 4.20: The height of $A_i V B_i$ cannot exceed $d_H(V, \mathcal{L})$.

Proof. In the model of the disk, we locate the right angle at the center of the disk by means of an isometry; in Figure 4.20 we show that the height cannot exceed the distance from the center of the disk to the line $\mathcal{L}$ through P_1 and P_2, where these are the points at infinity of the rays to which the congruent sides belong.

The reader must make the computation of such a constant. □

The fundamental property of hyperbolic geometry is the fact that through a point P not on a line $\mathcal{L}$, there exists more than one line which does not intersect $\mathcal{L}$, and we define as parallels only the lines sharing a point at infinity with $\mathcal{L}$; the remaining lines are called ultraparallels. The following result states how the angle formed by the parallels depends on the distance from P to $\mathcal{L}$.

5°. (Janos Bolyai, (1802–1860)). *The angle 2α formed by the two parallels to a line $\mathcal{L}$ through a point P not on $\mathcal{L}$ depends only on the distance from the point to the line (the* **angle of parallelism***).*

Proof. If we use the model of the disk, by applying an isometry to the point P and the line $\mathcal{L}$ so that it maps the point P into the center of the disk, it turns out that the angle 2α between the parallels to $\mathcal{L}$ is the angle between the radii determined by the points at infinity P_1 and P_2 of $\mathcal{L}$ (see Figure 4.21).

It is easy to see that the distance a from P to $\mathcal{L}$ increases as the arc $P_1 P_2$ closes; actually, the relation between 2α and a is

$$\cosh a \sin \alpha = 1. \tag{4.15}$$

The proof of this formula can be found in [Be], as well as several other results in hyperbolic trigonometry. □

The last result we mention in this section shows how the model of the upper half plane can be extended to higher dimensions, and it has the feature of

4.5. First results in hyperbolic geometry   181

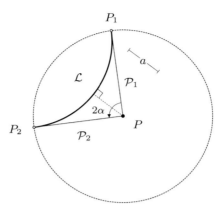

Figure 4.21: The angle of parallelism depends on a.

emphasizing the fact that in the hyperbolic space there are subsets (of smaller dimension), for which Euclidean geometry is valid.

We can take as a model for the 3-dimensional hyperbolic space the upper half space $H^+ = \{(x, y, z) \mid z > 0\}$ in $\mathbb{R}^3$, equipped with the scalar product defined similarly as for the plane: we modify the usual one dividing it by z^2, the square of the height above the XY plane (see Figure 4.22). The hyperbolic metric is then defined exactly as before. Analogously to the horocycles, a **horosphere** in the 3-dimensional hyperbolic space is a 2-plane contained in the upper half space, tangent to the plane of the points at infinity (see Figure 4.22). When adding to it the "point at infinity", this becomes a 2-sphere. This picture is clearer in the disk model for the 3-dimensional hyperbolic space Δ, where the horospheres are actual Euclidean spheres embedded in Δ so that they are tangent to the boundary of Δ at one point. Of course these concepts obviously extend to higher dimensions: in the n-dimensional hyperbolic space $\mathbb{H}^n$ one has horocycles and horospheres of all dimensions up to $n-1$.

6°. (Nikolai Ivanovitch Lobachevsky, (1793–1856)). *Euclidean geometry is valid in a sphere of infinite radius (i.e., a horosphere).*

Proof. We restrict to $\mathbb{H}^3$ for simplicity. The points of the XY plane are points at infinity and the geodesics are semi-circles perpendicular to the XY plane with its center at it, plus the half lines perpendicular to the XY plane, all of which have a common point at infinity that must be added to those of the XY plane. □

As stated in the result by Lobachevsky, in the planes parallel to the XY plane (horospheres that have this last point at infinity in common), where z is constant, the scalar product of vectors tangent to that plane differs from the Euclidean scalar product only in one constant: $1/z^2$. It is just like having the usual Euclidean space, but all distances are divided by this constant.

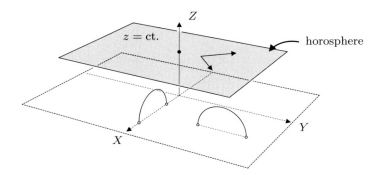

Figure 4.22: Model of the upper half space for the 3-dimensional hyperbolic space.

That is why in a horosphere the Euclidean geometry is valid.

It is worthwhile mentioning that, just as both Poincaré models, the Beltrami model generalizes easily to higher dimensions (see [Ve]). The model of the hyperboloid also generalizes to dimensions greater than 2 and is the most adequate model for constructing subgroups of isometries of the hyperbolic space H^n, with properties similar to those for the case $n = 2$ that we describe in the following sections .

To conclude this first approach to hyperbolic geometry, in the following section we show how to construct new objects whose geometry is hyperbolic, and in the last section of this chapter we deal with the problem of covering the hyperbolic plane with congruent tiles.

Exercises

1. Prove that any triangle in H^+ whose vertices are hyperbolic points can be obtained from triangles with an ideal vertex.

2. Compute the constant defined in [4].

3. Prove that two lines are ultraparallel if and only if they have a common perpendicular.

4. Obtain and justify a formula for the area of a hyperbolic polygon of n sides.

5. Prove that if $n \geq 3$, there exist regular polygons whose angles satisfy $0 < \alpha < (n-2)\pi/n$.

6. Justify calling an elliptic transformation a **rotation**.

7. In **4º**, prove that the hyperbolic line AD forms with OA and OD equal angles if and only if the center of the Euclidean circle that contains it belongs to the linear bisector of the angle AOD.

8. Compute the quotient of the hyperbolic area of a disk and the hyperbolic length of its boundary, and determine the limit of that quotient as the hyperbolic length of the radius approaches ∞. Compare your answer with that obtained in the Euclidean case.

9. Consider the model of the upper half space mentioned in result **6°**. What can be said about a half sphere with center on the XY plane?

4.6 Surfaces with hyperbolic structure

In Chapters 1 and 3, we created new geometric objects departing from other objects that we already knew well, by means of the technique of **forming the quotient space, modulo a subgroup** of the group of isometries under study; that is, the new object consists of the equivalence classes defined by the subgroup of the group of isometries under study.

Thus, by considering subgroups of translations of $\mathbb{R}^2$, first with only one generator, $T_{(1,0)}$ and then with two generators, $T_{(1,0)}$ and $T_{(1,0)}$, we obtained a cylinder and a torus $S^1 \times S^1$, respectively, which inherited the measure $\mathbb{R}^2$.

In Chapter 3, we obtained the elliptic plane by forming the classes defined in the sphere S^2 by the subgroup of isometries of the sphere, which has only two elements: the identity element and the antipode.

Now we shall introduce the theory equivalent to these in the hyperbolic case.

It is worth mentioning that this is a very rich and widely studied area of mathematics, where hyperbolic geometry (see, for example, [Ve]) and complex variables (more specifically, the theory of Riemann surfaces) converge (see [Spr] or [F]).

The whole theory can also be developed from the point of view of algebra and algebraic geometry, obtaining thus the algebraic curves. The required mathematics for those themes is beyond the purpose of this book, so we content ourselves with giving an intuitive introduction to this fascinating area of geometry.

Let us start by noticing that we know that for translations of $\mathbb{R}^2$ as well as for the antipode mapping in S^2, each point has a neighborhood in which there is only one representative element of the equivalence classes of each of its points (see Figure 4.23), and also in both cases the maps in question are isometries of the corresponding surface. That is why, in regions sufficiently small, the new object "behaves" like the original one from the metric viewpoint.

A subgroup of a group of isometries with the previous property is called a **discontinuous subgroup.**

That does not happen with a subgroup generated by a rotation or by a reflection.

If we consider a rotation by $2\pi/n$ in $\mathbb{R}^2$, any neighborhood of the point about which the rotation takes place contains several representative elements of the points that appear in it, and the same happens in the case of the sphere

184   Chapter 4. Hyperbolic geometry

S^2 when we consider a rotation of $\mathbb{R}^3$: the rotation leaves its axis fixed and for the points at which that axis intersects the sphere, it is not possible to find a neighborhood containing only one representative element of the points that appear in it. Something analogous happens if we consider the subgroup that generates a reflection, $\{I, Re\}$.

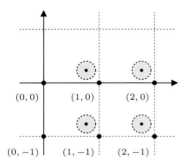 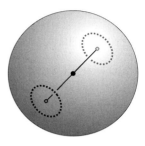

Figure 4.23: Each representative element has a neighborhood where there is only one representative element of the classes that appear in it.

In the case of the subgroup generated by a rotation, we would have a peculiar situation when forming the quotient object: a neighborhood of the fixed point is more like a cone than a plane, for the tangent vectors of differentiable curves through that point do not form a plane. In the case of a reflection, the points of the line with respect to which the reflection takes place do not have neighborhoods homeomorphic to a disk, but to half a disk including the diameter; those types of situations demand a special treatment, for they are "manifolds with boundary" (see [G-R]).

However, one can have very interesting groups constructed from rotations and or reflections, but having more than one generator. In fact the whole group $SO(3)$ consists of rotations.

Similar constructions can also be made in hyperbolic geometry, to obtain groups of isometries with interesting **quotient spaces**, obtained by considering the corresponding equivalence classes in the hyperbolic plane.

In the case of the hyperbolic plane (think of H^+), there are also transformations that leave fixed hyperbolic points, the elliptic transformations. We know already that near its fixed point, an elliptic transformation behaves like a rotation, and we have just seen that this gives problems when passing to the equivalence classes. These problems can be dealt with easily, but this needs a treatment which is beyond the scope of this book. So we restrict ourselves to the case of transformations without ordinary fixed points.

A first example of a discontinuous subgroup is given by the subgroup generated by a hyperbolic transformation T. That is, we consider T and all its iterates $(T \circ T)$, $(T \circ T \circ T)$, etc., as well as its inverses T^{-1}, $(T^{-1} \circ T^{-1})$, etc.

4.6. Surfaces with hyperbolic structure

Let us remember that a hyperbolic transformation T has its two fixed points at infinity, z^* and w^*, and therefore the geodesic line $\mathcal{L}$ they determine is invariant; hence, any point $P \in \mathcal{L}$ is transformed into another point $T(P) \in \mathcal{L}$, and the oriented distance between P and $T(P)$ does not depend on P, as the reader must prove using the standard form of f for the case of H^+: $f(z) = kz$, $(k > 0)$, whose fixed points are 0 and ∞ and, accordingly, $\mathcal{L}$ is the imaginary axis.

Actually, the effect of T on any point $R \in H^+$ can be obtained as follows: let us draw the perpendicular from R to $\mathcal{L}$, which we have taken as a diameter, and let us call H the foot of that perpendicular. We then apply T to H, and $T(R)$ is obtained by locating the point R' of the perpendicular to $\mathcal{L}$ through $T(H)$ such that $d(T(H), R') = d(H, R)$ (see Figure 4.24). Notice that this distance is an oriented distance because T respects the orientation of angles.

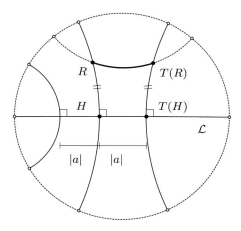

Figure 4.24: How to locate $T(R)$ for a hyperbolic transformation.

Given the special role played by the line $\mathcal{L}$ for determining the transformed point of any point R of the hyperbolic plane under T, we say that $\mathcal{L}$ is **the axis** of T, and the distance between $P \in \mathcal{L}$ and $T(P) \in \mathcal{L}$ is called the **distance of translation.**

The fact we wanted to prove is that the distance between a point and its transformed point is greater for points not on the axis of translation than in the case of points on $\mathcal{L}$, for the arc of the geodesic determined by R and $T(R)$, which from the Euclidean approach already has a length greater than that of the arc of $\mathcal{L}$, has a length even greater in the hyperbolic metric. Therefore, the distance between a point R and its point $T(R)$ transformed under a hyperbolic transformation takes its minimum at the points of the axis $\mathcal{L}$.

The subgroup of G_Δ generated by a single hyperbolic transformation T has as a fundamental region any stripe $\mathcal{D}$ bounded by two perpendiculars to the axis $\mathcal{L}$ whose feet are a distance $|a|$ apart. That is, if we consider the equivalence class in

Figure 4.25: The quotient of Δ by the subgroup generated by a hyperbolic transformation is a topologic cylinder.

H^+ given by $z \sim z'$ if and only if there exists $g \in G_\Delta$ such that $g(z) = z'$, then each equivalence class has a representative in such a strip $\mathcal{D}$, and this representative is unique for points in the interior of the strip.

The quotient of Δ by the subgroup generated by T looks like a cylinder, since the points at the edges of the strip are related by T in the way described above, and therefore they must be identified to another one (see Figure 4.25). Notice that in this hyperbolic cylinder there is a closed curve of minimum length.

In the case of Euclidean geometry, after constructing a cylinder as the set of equivalence classes defined by the subgroup generated by a single translation, we proceeded the same way with the subgroup generated by two translations, and we obtained a torus.

We would like to make a similar construction in hyperbolic geometry, to form the quotient of Δ by the subgroup generated by two hyperbolic transformations, T_1 and T_2, with different axes, for instance the lines $\mathcal{L}$ and $\mathcal{M}$ of Figure 4.26.

To obtain a fundamental domain corresponding to T_1, a strip as above, we can apply T_1 to $\mathcal{M}$; and to obtain a fundamental domain for T_2, we can apply T_2 to $\mathcal{L}$. The intersection of the two strips, that is, the quadrilateral which is the intersection of the fundamental domains of T_1 and T_2, in general does not have the desired properties, for the opposite sides do not have the same length (see Figure 4.26): $|BC|_\Delta < |AD|_\Delta$, and $|AB|_\Delta < |DC|_\Delta$, so we can not glue them by a hyperbolic isometry.

The only case where the opposite sides are congruent is when they have infinite length, as in Figure 4.27.

It is possible to cover the disk with quadrilaterals obtained from the shaded quadrilateral $\alpha\beta\gamma\delta$ under the hyperbolic transformations generated by T_1 with axis $\mathcal{L}$ and distance of translation $|a|$, and T_2 with axis $\mathcal{M}$ and distance of translation $|b|$.

To convince the reader, we suggest considering first the shaded quadrilateral and its infinite images under the hyperbolic transformation that leaves the vertical

4.6. Surfaces with hyperbolic structure

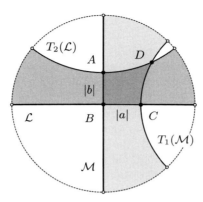

Figure 4.26: The opposite sides of the quadrilateral $ABCD$ are congruent only if the quadrilateral is ideal.

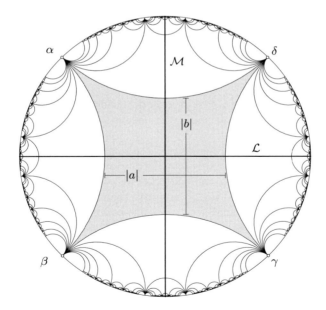

Figure 4.27: The images of the quadrilateral $\alpha\beta\gamma\delta$ under the subgroup generated by T_1 and T_2 cover Δ.

diameter fixed; the reader will agree that when the hyperbolic transformation that leaves the horizontal diameter fixed is applied, we not only obtain the image of the shaded quadrilateral $\alpha\beta\gamma\delta$, but also those of its images under the first transformation.

However, the object obtained by forming the classes of equivalence of Δ module the subgroup generated by T_1 and T_2 is not a topological torus because it is missing a point, for all the vertices, which are identified at a single point under the action of the group, are points at infinity. In complex geometry, that point is called a **puncture** of the quotient surface.

The previous discussion shows that we can obtain the punctured torus as quotient of the hyperbolic plane by the group generated by two hyperbolic isometries, with no fixed points in H^+. This means that the punctured torus admits **a hyperbolic structure**. If we could obtain the torus by a similar method, we would have that the torus admits a hyperbolic structure, and it is known that this is false, as the previous discussion suggests. We refer to [T] for a proof of this claim, which can also be found in many other books, together with many other fantastic results about this beautiful subject (see also the discussion below concerning Riemann's mapping theorem).

Here we are only concerned with groups that give rise to **compact surfaces**, those which are not missing a point (no punctures) and which do not have a boundary.

The transformations we are considering, $f \in PSL(2, \mathbb{C})$, are functions $\mathbb{C}$-differentiable at every point z_0 of their domain, that is, for which there exists the limit
$$\lim_{h \to 0} \frac{f(z_0 + h) - f(z_0)}{h}.$$
This limit is denoted as $f'(z_0)$; it is called the **derivative of** f at z_0 and is a complex number. The $\mathbb{C}$-differentiable transformations are also called **holomorphic or analytic** (see [A]).

Hence, two representative elements of the same point will have coordinates z_1 and z_2 related by a $\mathbb{C}$-differentiable transformation with $\mathbb{C}$-differentiable inverse.

At the end of section 3.4, we said that the surfaces with that property are called Riemann surfaces, and it is easy to prove that they turn out to be orientable. Accordingly, the objects we shall deal with are compact and orientable surfaces.

The surfaces with those properties have been widely studied, and we know what they are like because of two very important theorems. We do not include here the proofs of these theorems, but we do provide the reader the bibliographic references where they can be found.

The first theorem we shall mention belongs to the branch of mathematics called topology, and [I] is devoted to its proof; [I] is a very accessible book, although the proof also appears in the references about Riemann surfaces, for instance [Spr].

Theorem of classification of surfaces. *The only compact and orientable surfaces are the sphere and spheres with a finite number of handles (see Figure 4.28).*

4.6. Surfaces with hyperbolic structure

The number of handles the surface has is called its **genus**. A surface of genus 1 is called **a torus**, and it is homeomorphic to the product $S^1 \times S^1$. Between the genus and the Euler characteristic of the surface, $\mathcal{X}(S)$, there exists the relation (see [I] or [Spr]):
$$\mathcal{X}(S) = 2 - 2g.$$

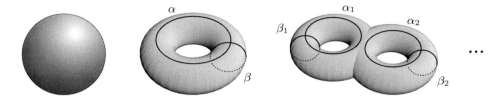

Figure 4.28: The compact and orientable surfaces are the sphere and spheres with handles.

Notice that each handle gives rise to two closed curves, α_i, β_i, which cannot be deformed into a point without leaving the surface, and it can be proven that it is not possible to deform any of those curves corresponding to a handle into another of them corresponding to another handle.

Topologists construct a surface of genus g from a polygon with $4g$ sides, whose sides they identify as shown by the arrows in Figure 4.29 for the case of genus 2.

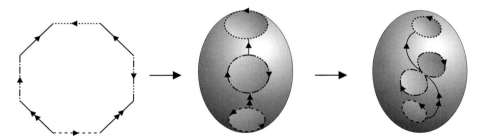

Figure 4.29: Topological construction of a torus with 2 handles.

We ask the reader to prove that on the hyperbolic plane there are regular polygons of every number of sides, greater than 4, whose angles add up to 2π. This assures the existence of elements in G_Δ which identify two sides with the same symbol and opposite orientations (as occurred with the translations in the case of a Euclidean square). This is also a consequence of a very important theorem, known as the **Uniformization Theorem**.

This theorem belongs to the branch of mathematics called complex geometry; it has been studied by many of the world's top mathematicians, and they have developed new theories to understand it fully. The proof of the Uniformization Theorem, which is beyond the scope of this book, was done simultaneously and independently in 1907 by Paul Koebe (1882–1945) and Henri Poincaré, but its origin is found in a famous theorem due to Riemann.

Theorem (Riemann's mapping). *For any open, proper, connected and simply connected subset (i.e., with no holes) R of $\mathbb{C}$, there exists a mapping $f : R \to \Delta$ which is a biholomorphism, that is, $\mathbb{C}$-differentiable with $\mathbb{C}$-differentiable inverse.*

Notice that in this dimension, a biholomorphism respects angles, including their orientation (why?). This justifies saying that if there is a biholomorphism between two surfaces, the surfaces are **conformally equivalent**.

We have already exhibited such a biholomorphism in the case when R is a half plane: the mapping $Q : H^+ \to \Delta$ given by (4.2).

The references for the theorem of uniformization are [Spr] and [F], and for Riemann's mapping theorem see [A].

For the compact Riemann surfaces we are dealing with, the usual version of the theorem of uniformization asserts that $\widehat{\mathbb{C}}$ has a (doubly) elliptic geometry, that the tori have a parabolic (plane) geometry, and that the surfaces of genus $g \geq 2$ have a hyperbolic geometry.

That is a consequence of the impossibility of establishing a biholomorphism f between the simplest Riemann surfaces from the topological point of view, $\widehat{\mathbb{C}}$, $\mathbb{C}$, and Δ (notice that biholomorphism implies homeomorphism):

- A sphere and a plane are not homeomorphic, for the sphere is compact and $\mathbb{C}$ is not (see [DoC], for characterizations of compact sets). That is why it is said that $\widehat{\mathbb{C}}$ and $\mathbb{C}$ are not conformally equivalent.

- In addition, although $\mathbb{C}$ and Δ are homeomorphic, it is not possible to establish a biholomorphism between them, for a $\mathbb{C}$-differentiable function whose domain is $\mathbb{C}$ and whose image is contained in the disk ΔD, is a bounded function; for that type of function, we have a theorem of complex variables, the theorem of Liouville (see [A]), which asserts that the function must be constant. Then the image of the function cannot cover Δ, and that is why $\mathbb{C}$ and Δ are neither conformally equivalent.

The formal statement of the theorem of uniformization is the following.

Theorem of uniformization. *Every Riemann surface can be endowed with an elliptic (or spherical) geometry, or with a parabolic (Euclidean) geometry, or with a hyperbolic geometry, depending on whether it is conformally equivalent to the set of equivalence classes of $\widehat{\mathbb{C}}$, $\mathbb{C}$ or Δ, these being determined by a group of isometries of the corresponding "plane" P whose elements have no fixed points in P.*

Here the plane P is the Riemann sphere in the first case, the Euclidean plane in the second, and the hyperbolic plane in the latter. The corresponding groups of isometries are $SO(3)$, $A(2)$ and $PSL(2, \mathbb{R})$.

4.6. Surfaces with hyperbolic structure

We remark that spherical geometry is also called (doubly) elliptic because the sphere is a double cover of the elliptic plane and the metric on the eliptic plane comes from that in the sphere. The only compact surface one has with this geometry is the Riemann sphere itself. The only compact parabolic surfaces are tori, but there are many such structures on tori which are not conformally equivalent, as many as points in $\mathbb{C}$ (see [A]). All other Riemann surfaces, i.e., all compact surfaces of genus $g > 1$, admit a hyperbolic geometry. In fact they admit many non-equivalent such geometries; these correspond bijectively with the points in a certain vector space (over $\mathbb{C}$) of dimension $3g - 3$, by a deep theorem of Riemann.

That is why we say that most of the compact and orientable surfaces admit a hyperbolic structure.

Now let us see how to construct a torus with two handles from a hyperbolic regular octagon whose angles are $2\pi/8$, when the sides are identified as shown in Figure 4.30. It is necessary to take care of the amplitude of the angles so that when identifying all the vertices at a point, the plane tangent to the quotient surface at that point is well defined. With that we guarantee that we have a well-defined metric in the new object, induced by the metric in Δ.

In the case of the Euclidean square, the identification of the opposite sides was made with translations by $T(1,0)$ and $T(0,1)$; in the case of the hyperbolic octagon, we identify the sides with the same symbol and opposite orientation by using the hyperbolic transformations h_i, whose axis is the perpendicular $\mathcal{H}_i$ common to the sides marked with the same symbol and whose distance of translation is that of the perpendicular segment between the sides. As the reader should have verified, that common perpendicular exists if and only if the lines are ultraparallels, as in this case.

The octagon is the fundamental domain of the subgroup G_2 whose direct isometries are generated by the four transformations h_1, h_2, h_3 and h_4, and it is clear that Δ can be covered with the images of the octagon illustrated in Figure 4.30, obtained from the elements of G_2.

Exercises

1. Prove that if T is a hyperbolic transformation with axis $\mathcal{L}$, the hyperbolic distance $d(P, T(P))$ does not depend on $P \in \mathcal{L}$.

2. Use the homeomorphism established by the function tangent between the interval $(-\pi/2, \pi/2)$ and $\mathbb{R}$ to give a homeomorphism between $\mathbb{C}$ and Δ.

3. Prove that every torus with g handles can be obtained from a regular hyperbolic $4g$-gon using hyperbolic transformations, if the angles maintain adequate amplitude.

4. What condition must a function $f : \mathbb{R}^2 \to \mathbb{R}$ satisfy so that it makes sense to say that f also defines a function from the Euclidean cylinder

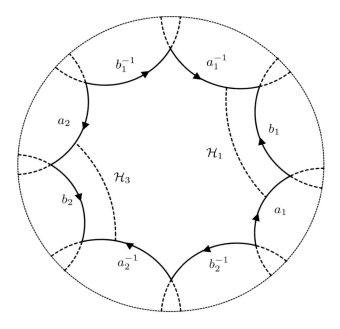

Figure 4.30: Geometric construction of a torus with 2 handles.

created as quotient of $\mathbb{R}^2$ by the subgroup generated by the translation $T_{(1,0)}$, with values in $\mathbb{R}$?

5. Establish the conditions that a function $f : \mathbb{R}^2 \to \mathbb{R}$ must satisfy for it to make sense to say that it also defines a function whose domain is the plane torus of Chapter 1.

6. Does it make sense to pose an analogous problem in the case of $f : \Delta \to \mathbb{C}$ and a torus with more than one handle?

7. Enlarge Figure 4.30 and add to it at least the octagons resulting from applying h_i and h_i^{-1} with $i = 1, 2, 3, 4$.

4.7 Tessellations

Now we envisage the problem of covering the hyperbolic plane with congruent tessellations, and finding the group of isometries that leave such a tiling invariant.

In contrast with the Euclidean and elliptic cases, there are infinitely many different ways of covering the hyperbolic plane with congruent tessellations; as we shall see, this is a consequence of the richness of hyperbolic geometry, where there is a great freedom to construct hyperbolic polygons which determine groups of

4.7. Tessellations

isometries with the polygon as fundamental domain. So its various images under the group cover the whole plane and give a tiling of it.

For example, let us consider triangles, say in 3-space, with angles $\frac{\pi}{p}, \frac{\pi}{q}, \frac{\pi}{r}$, $p, q, r \geq 2$, and order them so that $p \leq q \leq r$. There are three obvious possibilities, according as the sum $\frac{\pi}{p} + \frac{\pi}{q} + \frac{\pi}{r}$ is greater than, equal to, or smaller than π.

When $\frac{\pi}{p} + \frac{\pi}{q} + \frac{\pi}{r} = \pi$, it is well known that every such triangle T can be realized as a Euclidean triangle in $\mathbb{R}^2$. Each of its three edges determines a reflection of $\mathbb{R}^2$, and the set of all possible compositions of these three reflections forms a subgroup $\Sigma(p, q, r)$ of the group $A(2)$, called **a triangle group**. It is an exercise to see that the various images of T under the transformations in $\Sigma(p, q, r)$ fill out the whole plane and they are pairwise disjoint, except for points in the boundary of the triangles. We thus get a Euclidean tiling of the plane, as described in Chapter 1. It is clear that the only unordered triples (p, q, r) that satisfy these conditions are $(2, 3, 6)$, $(2, 4, 4)$ and $(3, 3, 3)$. In the later case the tiling is by equilateral triangles. The case $(2, 4, 4)$ corresponds to taking the tiling by squares and triangulating each square by joining with lines its center with the four vertices and with the middle-points of the edges. This triangulation of the square is called its **barycentric** subdivision (we leave as an exercise to show that the triangles one gets in this way are indeed $\frac{\pi}{2}, \frac{\pi}{4}, \frac{\pi}{4}$). The case $(2, 3, 6)$ is the tiling defined by an hexagon, taking its barycentric subdivision.

If we look at triangles with $\frac{\pi}{p} + \frac{\pi}{q} + \frac{\pi}{r} > \pi$, it is easy to see that the only possibilities for the unordered triple (p, q, r) are $(2, 2, r), r \geq 2$, $(2, 3, 3)$, $(2, 3, 4)$ and $(2, 3, 5)$. The triangle T is now a spherical triangle: it can be realized in the 2-sphere, bounded by geodesics.

It is well known that the last three cases correspond to the tilings determined by a tetrahedron, an octahedron (or equivalently, a cube) and a dodecahedron (or equivalently, an icosahedron), that we studied in Chapter 1. The triangles arise by taking the barycentric sub-division of the corresponding Platonic solid, and then projecting the solid into the sphere from the center. We refer, for instance, to Chapter II in [Se] for a careful description of this situation, whose study goes back to the work of F. Klein and others in the 19th Century. The corresponding group of isometries of the sphere that preserve this tiling is the subgroup of $O(3)$ generated by reflections on the planes in $\mathbb{R}^3$ determined by the edges of the triangle.

Notice that in all the above cases, the group of symmetries that preserves the tiling is generated by reflections on the sides of the triangle T, which is a fundamental domain. In both cases we may look at only the index 2 subgroup, of the corresponding triangle group, consisting of elements that can be expressed as composition of an even number of reflections. Since the composition of two reflections on lines which are not parallel defines a rotation, these are groups of rotations that preserve the corresponding tiling.

The case $\frac{\pi}{p} + \frac{\pi}{q} + \frac{\pi}{r} < \pi$ is somehow the "richest", for we now have infinitely many possibilities for p, q, r. Given any such triple, one has that the triangle T can be realized in the hyperbolic plane and one has the corresponding tilings, as we explain below.

As to the playful aspect of this problem, of constructing tilings of the hyperbolic plane, it was the already mentioned Dutch engraver M. C. Escher who used it magisterially, in spite of, at least at the beginning, not knowing that he was using hyperbolic transformations. [1]

Escher decorated the hyperbolic polygons with different images to create several of his most celebrated engravings; if the reader analyzes them from the point of view of this section (see [Er] and [Es]), he or she will gain hyperbolic intuition, for they illustrate how the decoration changes by rotating it about a point, by reflecting it with respect to a hyperbolic line, and by translating it on a line.

No doubt that one of the best and most joyful ways for gaining intuition of hyperbolic geometry is to create some tilings with the methods presented here.

Two of the figures in the previous section, 4.27 and 4.30, show how to cover the Poincaré disk with hyperbolic tessellations: the first ones are ideal quadrilaterals, with their vertices at infinity, the second ones would be regular octagons (they have not been drawn, but we asked the reader to do it in Exercise 7, Section 4.5) whose vertices are ordinary points and whose interior angles are $2\pi/8$. The octagons that have not been drawn are obtained by means of the hyperbolic transformations used to "glue" the sides with the same symbol and opposite orientations, for each of them translates the octagon along the perpendicular common to those sides; by using all the hyperbolic transformations generated by those transformations, we can cover Δ with regular octagons.

In both cases, the transformations used to cover the disk are transformations of hyperbolic type, which in a sense are analogous to the Euclidean translations. In the case of the ideal quadrilateral, two hyperbolic transformations suffice to generate the subgroup that covers ΔD, but in the case of the octagon with ordinary vertices, it was necessary to use four hyperbolic transformations to generate the group that covers Δ.

Thus, we already have an infinite number of ways of covering Δ with regular (and different) tessellations:

- With ideal polygons of $2k$ sides, with $k \geq 2$; the reader ought to describe the group of isometries.

- With regular $4n$-gons whose sum of angles is 2π; the group of isometries of that tiling is generated by the $2n$ hyperbolic transformations whose axes are the perpendiculars common to the sides that must be identified, and where the distances of translation of each is the distance between those sides (see Figure 4.27).

---

[1] "In mathematics I did not ever get a C. It is curious that, it seems, I have been occupied with mathematics without realizing it. Who could imagine that mathematicians would illustrate their books with my drawings, that I would hobnob with great scholars as if they were my colleagues and brothers!" Escher, in [Er].

4.7. Tessellations

We now want to return to discuss the triangle groups in hyperbolic geometry.

We use the disk model. The first step is to describe the transformations of Δ that play the role of reflections. These are known as **inversions**, and we discuss them briefly in Appendix 5.6.

On "the real line", the inversion with respect to the interval $[-1, 1]$ is the map $x \mapsto \frac{1}{x}$. Notice that this map carries the origin to ∞, so it has to be defined on the "extended line" $\hat{\mathbb{R}}$, which is homeomorphic to the circle S^1, as we know already. In this case we have that the map is taking outside the inside of the interval and vice-versa, with the end points $\{-1, 1\}$ being fixed. Of course a similar expression can be given in general, to define the inversion with respect to an arbitrary interval $[x_o - \epsilon, x_o + \epsilon]$; we leave as an exercise to find the corresponding formula.

Inversions extend to automorphisms of $\hat{\mathbb{C}}$ with respect to circles in the obvious way. If the circle has infinite radius, so it is a Euclidean line, the corresponding inversion is the usual Euclidean reflection. To define the inversion with respect to a circle $\mathcal{C}$ (of finite radius), we do it line by line, with respect to all lines passing by the center x_o of the circle. Every such line can be considered as a copy of $\mathbb{R}$, and it meets the disc bounded by $\mathcal{C}$ in an interval of radius that of $\mathcal{C}$, so we can make the inversion on this line just as above.

Analogously to the case of reflections in the Euclidean case, the formula to determine an inversion in general can be obtained by taking the circle $\mathcal{C}$ into the real line by a Möbius transformation T, then taking the reflection on the straight line $T(P)$, using the formula for the reflection on $\hat{\mathbb{R}}$ and then applying T^{-1}. We refer to our Appendix 5.6 for more about inversions.

Let us now look at the way inversions transform the disc Δ, which is equipped with the hyperbolic metric. We are actually interested in the inversions that preserve Δ, i.e., inversions ι in $\hat{\mathbb{C}}$ such that $\iota(\Delta) = \Delta$ as a set. For instance, if $\mathcal{C}$ is a straight line in $\mathbb{C}$ that passes through the center 0 of Δ, then the Euclidean reflection in $\mathcal{C}$, which is an inversion, obviously carries Δ into itself.

We ask the reader to show that inversions are conformal maps, i.e., they preserve angles, and therefore, given the disc Δ, regarded as a subset of $\mathbb{C}$, one has that the inversion on a given circle $\mathcal{C}$ carries Δ into itself if and only if $\mathcal{C}$ meets the boundary $\partial \Delta$ of Δ perpendicularly (prove this statement!). According to Section 2 of this chapter, we then have that **inversions on circles orthogonal to the boundary of Δ are isometries for the hyperbolic metric**. Moreover, the composition of an even number of inversions preserves the orientation, so every such map is a Möbius transformation, by Section 2.

If $\mathcal{C}$ is a circle in $\mathbb{C}$ that meets $\partial \Delta$ perpendicularly, denote by $\mathcal{L}$ the hyperbolic line that it defines, i.e., its intersection with Δ. Then we can speak of the inversion $\iota = \iota_{\mathcal{L}}$ of Δ defined by $\mathcal{L}$, which is a transformation of Δ into itself, and we already know that this is an isometry. Notice that, in analogy with the Euclidean reflections, given a point $P \in \Delta$, its image $P' = \iota(P)$ can be characterized by being the unique point in the hyperbolic line perpendicular to $\mathcal{L}$ at P, which is in the other half plane determined by $\mathcal{L}$, and whose distance to $\mathcal{L}$ equals the distance from P to this line (see Figure 4.31).

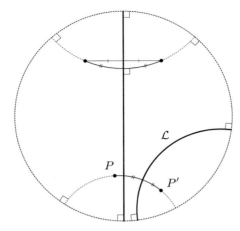

Figure 4.31: How to reflect on a hyperbolic straight line for the model of the disk.

Returning to triangles in the hyperbolic disc, we leave as an exercise for the reader to illustrate the form in which any ideal triangle and its reflected ones (and the reflected ones of the reflected ones, and so on) on each of its sides cover Δ.

Now let $T \subset \Delta$ be a hyperbolic triangle whose vertices are all ordinary points. In order for it to determine a tiling of Δ, its angles must satisfy that around each vertex V_i, a finite number $2k_i$ of images of T cover a neighborhood around that vertex *without* superimposing.

That implies, if the angle at V_i is α_i,

$$2k_i \alpha_i = 2\pi, \text{ that is, } \alpha_i = \frac{\pi}{k_i}. \tag{4.16}$$

The coefficient 2 appears because we are going to apply a reflection on each side of the triangle, and we want the resulting tiling to remain invariant under rotations about each vertex, as occurs in the Euclidean case, where the type of rotation was useful for classifying the tessellations.

An example will help the reader to take care of the problem that arises when the angle does **not** permit an even number of reflections. Let us take a triangle with one vertex at the center of the disk, and whose angle at that vertex is $\alpha = 2\pi/5$.

After marking one of the two sides contained in radii, if we proceed to reflect the triangle on those sides to cover the pentagon, we will not be able to obtain a pentagon invariant under rotations about the center (Exercise 3).

Accordingly, each angle of a hyperbolic triangle, being the basic piece to construct a hyperbolic tiling, must be of the form π/n, with $n \in \mathbb{N}$. This is why we take the angles to be $\frac{\pi}{p}, \frac{\pi}{q}, \frac{\pi}{r}$.

4.7. Tessellations

On the other hand, we know that the sum of the three angles must be less than π; hence, if $\alpha_1 = \pi/p$, $\alpha_2 = \pi/q$ and $\alpha_3 = \pi/r$, we obtain a restriction on the numbers $p, q, r \in \mathbb{N}$:

$$\frac{1}{p} + \frac{1}{q} + \frac{1}{r} < 1. \tag{4.17}$$

Now, given the triangle T with angles $\frac{\pi}{p}, \frac{\pi}{q}, \frac{\pi}{r}$, we let Σ be the group of isometries of Δ generated by the inversions on the three edges of T. Then the images of T by the elements of the triangle group Σ give a tiling of the hyperbolic plane.

A triple (p, q, r) which satisfies these conditions is $(7, 2, 3)$ and the corresponding tiling is given in Figure 4.32. This corresponds to the hyperbolic triangle of largest possible area (show it!).

Figure 4.32: Tessellation whose basic triangle corresponds to $(7, 2, 3)$.

The group of Möbius transformations that leave this tessellation invariant has as generators the rotation by $2\pi/7$ around the center of the central heptagon, the rotation by $2\pi/3$ around one of the vertices of the heptagon, and the rotation around the foot of an apothem of the heptagon (the perpendicular from the center to one side).

For readers interested in this topic, the best reference is [Be].

Exercises

1. Mark at random four points on the boundary of Δ and draw the hyperbolic quadrangle with those vertices; give a geometric algorithm, and justify it, to obtain the images of that quadrangle under the reflections on its sides.

2. In the model of the disk, draw a regular pentagon with center at $(0,0)$ and divide it into triangles as indicated above. Determine if it is possible with the marked triangle to generate a tiling whose group of isometries is non-trivial.

3. Consider the case of a triangle T such that not all of its vertices are either ideal or ordinary; draw it on both Poincaré models. Can we cover Δ with it? Justify your answer.

4. At least for three of the engravings by Escher that cover the hyperbolic plane, determine the values p, q, and r of the basic triangle, and the subgroup of G_Δ that leaves the tiling invariant. (Suggestion: see [Es].)

5. Design at least two hyperbolic tessellations.

6. Given an n-sided convex hyperbolic polygon $\mathcal{P}$ in Δ, can you find conditions to insure that the inversions on the edges of $\mathcal{P}$ define a group of transformations such that the various images of $\mathcal{P}$ give a tiling of Δ?

Appendices

5.1 Differentiable functions

To extend the concept of differentiability to a function of several variables, $F : \mathbb{R}^n \to \mathbb{R}$, at a point P_0 of its domain, we must recall that the value of a differentiable function at a point close to P_0, $P_0 + \bar{h}$, can be approximated by the value of the function at the point, $F(P_0)$, plus the value of a linear transformation, called precisely the **derivative at** P_0, applied to the increment $\bar{h}$, where $||\bar{h}||$ is sufficiently small.

In the case of a function $F : \mathbb{R}^3 \to \mathbb{R}$, the linear transformation that gives rise to the approximation can be thought of as a vector, called the **gradient** of F at P_0, $\nabla F(P_0)$, formed with the so-called **partial derivatives** of F at $P_0(x_0, y_0, z_0)$, which can be obtained as follows (we illustrate it with only one of them):

$$\frac{\partial F}{\partial x}(x_0, y_0, z_0) = lim_{h \to 0} \frac{1}{h} \left(F(x_0 + h, y_0, z_0) - F(x_0+, y_0, z_0) \right).$$

It is clear that the partial derivative with respect to one variable considers only the variation with respect to that variable, while the other variables remain constant.

Thus, $\nabla F(P_0)$ is formed as follows:

$$\nabla F(P_0) = \left(\frac{\partial F}{\partial x}(x_0, y_0, z_0), \frac{\partial F}{\partial y}(x_0, y_0, z_0), \frac{\partial F}{\partial z}(x_0, y_0, z_0) \right),$$

and is applied to a vector $\bar{u} = (h, k, l)$ by means of the scalar product,

$$\nabla F(P_0)(\bar{u}) = \left(\frac{\partial F}{\partial x}(x_0, y_0, z_0), \frac{\partial F}{\partial y}(x_0, y_0, z_0), \frac{\partial F}{\partial z}(x_0, y_0, z_0) \right) \cdot (h, k, l)$$

It can be proved that if each of the partial derivatives is continuous and $||(h, k, l)||$ is sufficiently small, the approximation

$$F(x_0 + h, y_0 + k, z_0 + l) \sim F(x_0, y_0, z_0) + \nabla F(P_0) \cdot (h, k, l),$$

is valid, in the sense that the difference between both members approaches zero more quickly than $||(h, k, l)||$.

For example, for $F(x,y,z) = x^2 + 4y^2 + 9z^2$, the partial derivative of F with respect to x at P_0, the partial derivative of F with respect to y at P_0, and the partial derivative of F with respect to z at P_0 are, respectively,

$$\frac{\partial F}{\partial x}(x_0, y_0, z_0) = 2x_0;$$
$$\frac{\partial F}{\partial y}(x_0, y_0, z_0) = 8y_0;$$
$$\frac{\partial F}{\partial z}(x_0, y_0, z_0) = 18z_0,$$

and the gradient of F at $(-1, 1, -1)$ is

$$\nabla F(-1, 1, -1) = (-2, 8, -18).$$

Of course the derivative of a function $f : \mathbb{R}^n \to \mathbb{R}$, when it exists, is not a vector, but a linear map defined at each point by the $1 \times n$ matrix

$$Df = (\frac{\partial F}{\partial x_1} \quad \frac{\partial F}{\partial x_2} \quad \cdots \quad \frac{\partial F}{\partial x_n}).$$

A value $r \in \mathbb{R}$ of F is a **regular value** if for every $(x_0, y_0, z_0) \in \mathbb{R}^3$ such that $F(x_0, y_0, z_0) = r$, the gradient of F at (x_0, y_0, z_0) is not zero.

The set of the points that are mapped into the same value r is called the **inverse image** of r (or the level surface of r) and is denoted as $F^{-1}(r)$. For instance, the sphere of radius 1 and centered at the origin is the inverse image of 1 under the function $F(x, y, z) = x^2 + y^2 + z^2$, and analogously for each of the quadric surfaces.

A point $(x_0, y_0, z_0) \in \mathbb{R}^3$ where the gradient $\nabla F(x_0, y_0, z_0)$ becomes zero is known as a **critical point** of F, and $r \in \mathbb{R}$ is a **critical value** of F if there exists a critical point in the inverse image of r.

In the case of the cone, which is an inverse image of 0 under the function $F(x, y, z) = x^2 + y^2 - z^2$, each of the generatrices is a curve on the cone that passes through the vertex, and that proves that the vectors tangent to curves at the vertex do not form a plane; that is why the plane tangent to the cone at the vertex cannot be defined, because $\bar{0} \in \mathbb{R}^3$ is a critical point (the only one) of the function $F(x, y, z) = x^2 + y^2 - z^2$.

Each point of $\mathbb{R}^3$ is contained in a level surface of this function, and that is why it is said that the level surfaces of this function **foliate** the space, they fill it with their **leaves**. The name comes from the Latin *folia*, which means leaf. We notice, however, that the origin plays a special role, because at this point the gradient of F vanishes.

The Implicit Function Theorem, which can be found in many text-books, grants us that away from $\bar{0} = (0, 0, 0)$ all the level surfaces of F are smooth, which intuitively means that the corresponding surface has a well-defined tangent space at the point. And the level surface $F^{-1}(0)$ is not smooth at $\bar{0}$. This is why the foliation defined by F is, properly speaking, **a singular foliation**, defining an actual foliation on $\mathbb{R}^3 \setminus \{\bar{0}\}$.

Notice that there is a qualitative change in the type of leaves when passing from $r < 0$ to $r > 0$; these surfaces are all smooth, and the change from one type of hyperboloid to the other passes through the cone, which is the level surface 0. The cone divides the two types of hyperboloids, and that is why it is called a **separatrix** of the foliation. To further study these themes, see [Hr].

5.2 Equivalence relations

In a set A, we have an **equivalence relation** if we have determined a relation R among its elements such that it satisfies the three following conditions:

i) Reflexivity: Any $a \in A$ is related to itself: aRa.

ii) Symmetry: If $a \in A$ is related to $b \in A$, then b is related to a: $aRb \Rightarrow bRa$.

iii) Transitivity: If $a \in A$ is related to $b \in A$ and b is related to $c \in A$, then a is related to c: aRb and $bRc \Rightarrow aRc$.

The elements related to one another form a class of equivalence, and two different classes cannot have common elements, because of transitivity. Hence, it is said that

An equivalence relation in a set A induces a partition of A into disjoint classes.

The set of equivalence classes is called the **quotient set of A modulo the equivalence relation**.

There are ordinary examples of equivalence relations. For instance, among the days of a month, the name of a day is the same as that of another if their numbers differ by a multiple of 7: if day 3 is Sunday, the other Sundays are the days 10, 17, and 24.

Among the integer numbers, a relation very frequently used comes from the residue left by those numbers when divided by a fixed number, 5 for example: the class of 0 is formed by the numbers divisible by 5; the class of 1 is formed by the numbers of the form $1 + 5k$, for any $k' \in \mathbb{Z}$; the class of 2 is formed by the numbers of the form $2 + 5k$, and so on. Each number has its **residual class** modulo 5 well-defined.

A function f between two sets, $f : A \to B$, with domain A, induces an equivalence relation in A when considering that two elements $a, a' \in A$ are related if $f(a) = f(a')$. The equivalence classes are the subsets of A which are the **inverse image** of a fixed element $b \in B$.

In the particular case of a function $f : \mathbb{R}^n \to \mathbb{R}$, the inverse image of a value $r \in \mathbb{R}$ is called the **level set** r of the function. If $n = 2$, the level sets are **level curves**; if $n = 3$, they are **level surfaces**, and so on.

5.3 The symmetric group in four symbols: S_4

The bijections of a set into itself can be seen as rearrangements of the elements of the set, and thus, when the set has a finite number of elements, a bijection is called a **permutation**.

The symbols 1, 2, 3, 4 can be arranged in 4! different ways, because the first symbol can be chosen among 4 elements, for the second symbol there are only 3 options, for the third, only 2, and for the last symbol there remains only one option.

Since 4! = 24, it is worthwhile writing all the arrangements; we first write those beginning with 1, then those beginning with 2, and so on. They are

$$\begin{matrix} 1234 & 1243 & 1324 & 1342 \\ 2134 & 2143 & 2314 & 2341 \\ 3124 & 3142 & 3214 & 3241 \\ 4123 & 4132 & 4213 & 4231 \end{matrix}.$$

Any arrangement can be obtained from another by a permutation, and to determine it, we write one arrangement under another, and we say that the first symbol of the second arrangement has substituted that which appears in the first place in the first arrangement. In the first arrangement we then look for the symbol that substituted the first one, and we see what symbol of the second arrangement is in the same place, and so on.

For example, if we take 1234 and 3124, we write

$$\begin{pmatrix} 1 & 2 & 3 & 4 \\ 3 & 1 & 2 & 4 \end{pmatrix},$$

and we read "1 has been replaced by 3, 3 has been replaced by 2, and 2 has been replaced by 1." What we briefly write is in the form of a **cycle**:

$$(132),$$

which must be interpreted as follows: 1 is replaced by 3, 3 is replaced by 2, and 2 is replaced by 1.

Notice that 4 remained in its original place, and that is why it is not necessary to write it.

When the permutation is reduced to an exchange of two symbols, as occurs with the arrangements 4132 and 4231, the cycle consists only of those two symbols, (12), and it is called a **transposition**.

Permutations can be composed, for they are functions, and the result of applying one permutation first and then another is a permutation of the symbols as well because the composition of two bijections is a bijection. Since the composition of functions is associative, and the inverse of a bijection is another bijection, it turns out that the permutations of four symbols form a group called the **symmetric group** in four symbols, which is denoted as S_4.

5.3. The symmetric group in four symbols: S_4

The identity element of this group is the permutation where every symbol remains in its place, and it is written as a product of cycles of length 1: (4)(3)(2)(1).

To determine the permutation that results from applying first the permutation (234) and then the permutation (13), we write the arrangement 1234 and we apply the two permutations following the proposed order, to the right (234) and to the left (13), as is usual in the composition of functions, (13)(234). In the next rows we can see the effect of applying the permutation (234) to the elements of the first row and then the permutation (13) to the elements of the second row:

$$\begin{pmatrix} 1 & 2 & 3 & 4 \\ 1 & 3 & 4 & 2 \\ 3 & 1 & 4 & 2 \end{pmatrix}.$$

The comparison of the last row with the first one reveals that 1 is replaced by 3; 3 is replaced by 4; 4 is replaced by 2, and 2 is replaced by 1. That permutation is represented by the cycle (1342), (remember that the order in the first member of the equality corresponds to that of the composition of functions) and thus the product of the permutation (234) and the permutation (13) is

$$(13)(234) = (1342).$$

It is clear that $(1342) = (3421) = (4213) = (2134)$, since all these permutations have the same effect on any arrangement.

The cycles are called **2-cycles**, **3-cycles** or **4-cycles**, depending on the number of symbols appearing in it. Notice that the product of two disjoint transpositions cannot be written as a 3-cycle, neither as a 4-cycle, as is the case of (23)(14).

The number of permutations of 4 symbols is $4! = 24$, for each arrangement is the result of applying one permutation to the "natural" arrangement 1234, and the number of arrangements is 24.

Another way of counting is as follows: since each permutation has an expression in cycles, we can count the number of 2-cycles, that of 3-cycles, that of 4-cycles, and that of products of disjoint 2-cycles, and then add them up together with the identity permutation (which can be seen as a product of 1-cycles):

- The number of 2-cycles is $\frac{4 \cdot 3}{2} = 6$.
- The number of 3-cycles is $\frac{4 \cdot 3 \cdot 2}{3} = 8$.
- The number of 4-cycles is $\frac{4 \cdot 3 \cdot 2 \cdot 1}{4} = 6$.
- The number of products of two disjoint 2-cycles is $6/2 = 3$, because since they are disjoint, the order of the factors does not affect the product.

If we add to these $6 + 8 + 6 + 3 = 23$ permutations the identity permutation, we have 24 permutations.

The theme is far from being exhausted in this introduction (we suggest the reader see [Bi]), but we hope to convince the reader of its importance by obtaining all the isometries of the tetrahedron using the elements of S_4S4.

We only need a fundamental fact which is not difficult to prove:

Remark. Any permutation of n symbols can be obtained as a product of a finite number of transpositions (not necessarily disjoint).

For example, the permutation (243) can be obtained as (verify it):

$$(243) = (24)(23).$$

It is easy to prove that any 4-cycle is a product of three transpositions, and that any 3-cycle is a product of two transpositions.

Because of the remark, it turns out that

1. The transpositions generate the symmetric group, and the permutations can be classified as even or odd, depending on the number of transpositions necessary to obtain them.

   The subgroup of the even permutations is called the **alternate group**, and in the case of four symbols is denoted as A_4.

   With respect to the isometries of the tetrahedron, let us notice two facts:

2. Any isometry of the tetrahedron gives rise to a permutation of its vertices, that is, it corresponds to an element of S_4.

3. In the tetrahedron, any two vertices are the same distance apart: the length of one edge (that does not occur in the cube).

Now we are ready to begin the analysis of the isometries of the tetrahedron.

- For any transposition, for example (23), there is an isometry that exchanges those vertices leaving the other two fixed, 1 and 4: the reflection on the plane containing the fixed vertices (1 and 4) and the midpoint M_{23} of the edge of the other two. For four symbols, there are six transpositions, which matches with the fact that there are six midpoints of edges.

The product of two transpositions corresponds to the composition of two reflections:

- If the transpositions are disjoint, like (23) and (14), the two associated reflections leave the midpoints M_{23} and M_{14} fixed. Then the straight line they determine remains fixed, and the isometry is a rotation by 180° about that axis. We know there are three permutations which are products of disjoint cycles, and that corresponds to the fact that there are three pairs of opposite edges.

- If the transpositions are not disjoint, like (12) and (13), the product is a 3-cycle, $(13)(12) = (123)$, which leaves 4 fixed. The corresponding isometry is a rotation by 120° about the straight line through the vertex 4 and through the centroid of the face 123, B_{123}. The product in the opposite order, $(12)(13) = (132)$, is associated to the rotation about the same axis by 240°. There are eightof these 3-cycles, corresponding to the fact that there are four faces and two rotations, which are not the identity, for each.

We just need to find the isometries associated to 4-cycles, like (1324). These cycles are products of three transpositions, in this case, $(1324) = (14)(23)(12)$. By the associativity of the composition of functions, we can multiply first the last two, $(23)(12) = (132)$, which produces the rotation mentioned at the end of the previous paragraph; that rotation follows from the reflection associated to the transposition (14). No vertex remains fixed and, of course, the orientation does not change: Compare how to go from 1 to 2 and then to 3 on the original face and on the final face containing those vertices. We know that the number of 4-cycles is six.

In this case the geometric justification of the number of those isometries comes directly from the exchange between the vertices, none of which remains fixed under a 4-cycle. Once the vertex i into which we shall apply the vertex 1 has been chosen (there are only three possibilities), the vertex i can be applied only into two vertices, for it cannot be applied into 1 because we would have a 2-cycle and, necessarily, another (and only another one) so that all the vertices would be exchanged, which is not the case.

We only need to add to these 23 isometries the identity (which is written as a product of 1-cycles: $(1)(2)(3)(4)$) to obtain the 24 possible isometries of the tetrahedron.

5.4 Euclidean postulates

The original form of the Euclidean postulates is as follows (see[E]).
   Let the following be postulated:

  I. To draw a straight line from any point to any point.

 II. To produce a finite straight line continuously in a straight line.

III. To describe a circle with any center and distance.

 IV. That all right angles are equal to one another.

  V. That, if a straight line falling on two straight lines makes the interior angles on the same side less than two right angles, the two straight lines, if produced indefinitely, meet on that same side on which are the angles less than the two right angles.

Today, Postulate V is presented at schools as **Playfair's Axiom**. It was John Playfair who used this equivalent form of Postulate V in his attempt to prove it on the basis of postulates I to IV.

Playfair's Axiom is used in this book to establish the two possible denials of Postulate V:

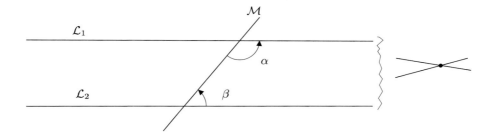

Figure 5.1: If $\alpha + \beta < 180°$, $\mathcal{L}_1$ and $\mathcal{L}_2$ intersect on that side of $\mathcal{M}$.

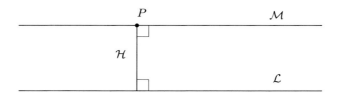

Figure 5.2: $\mathcal{M}$ is the only parallel to $\mathcal{L}$ through P.

Playfair's Axiom. *Through a given point P exterior to one straight line $\mathcal{L}$, only one parallel can be drawn to $\mathcal{L}$.* (See Figure 5.2.)

Since Playfair's Axiom establishes both the existence and the uniqueness of a parallel to a straight line $\mathcal{L}$ through a given point P exterior to $\mathcal{L}$, there are two forms of denying Postulate V:

N1 Through a given point P exterior to a given straight line $\mathcal{L}$, it is not possible to draw any parallel to $\mathcal{L}$.

N2 Through a given point P exterior to a given straight line $\mathcal{L}$, it is possible to draw more than one parallel to $\mathcal{L}$.

5.5 Topology

In calculus courses, the notion of an **open set** of $\mathbb{R}^n$ is described as a set A such that any of its points is the center of a ball of positive radius completely contained in A.

It is easy to verify that the open sets of $\mathbb{R}^n$ have the following properties:

1. The empty set and the whole ($\mathbb{R}^n$) are open sets.
2. The intersection of two open sets is an open set.
3. The arbitrary union of open sets is an open set.

5.5. Topology

The definition of a **topological space** establishes precisely these conditions. It is said that a set E is a topological space if among the subsets of E there is a family $\mathcal{F}$, the **family of the open sets of** E, which satisfies the three conditions given above. The family $\mathcal{F}$ is a **topology** for E, and each open set is a **neighborhood** of each of its points.

Notice that the definition of a continuous function from $\mathbb{R}$ into $\mathbb{R}$ can be read as:

The function $f : \mathbb{R} \to \mathbb{R}$ is **continuous at a point** x_0 of its domain if given an open set B containing $f(x_0)$ there exists an open set A containing x_0 and such that $f(A) \subset B$.

Therefore, it is possible to define continuous functions between any two topological spaces E_1 and E_2: it suffices to require that the inverse image of any open set of E_2 under f be an open set of E_1.

If a function f is continuous and bijective, there exists its inverse function, f^{-1}, and if this function is also continuous, it is said that E_1 and $f(E_1) \subset E_2$ are homeomorphic (do not confuse with homomorphic, a notion that belongs to algebra), and that f is a **homeomorphism**.

It is very important for readers to ask themselves which of the functions they know are continuous, and among these which are homeomorphisms. For instance, the projection $\Pi : \mathbb{R}^2 \to \mathbb{R}$ such that $\Pi(x, y) = x$ is a continuous function when we consider the usual open sets in $\mathbb{R}^2$ and those in $\mathbb{R}$, but it is not a homeomorphism because it is not injective.

When we have a topology defined in a set E, it is possible to make any of its subsets, E', become a topological space by endowing it with the induced (or relative) topology, which means that $A' \subset E'$ is a **relative open set** of E' if there exists an open set A of E whose intersection with E' is A', that is, $A' = A \cap E$. The reader will have no difficulty in verifying that the relative open sets satisfy the three conditions of a family of open sets.

Now it is easy to prove that the stereographic projection of a circle on the real straight line to which we add a point (see Figure 3.3) is a homeomorphism.

The open sets of S^1 will be the relative open sets of S^1 as subsets of $\mathbb{R}^2$, and the open sets of $\mathbb{R} \cup P_\infty$ are the usual neighborhoods of points of the straight line and the neighborhoods of P_∞ which we define as the complements of **bounded sets**, that is, sets that can be enclosed in an interval $(-R, R)$ with $0 < R < \infty$.

Again, the best the reader can do is to verify that by adding these open sets to the usual open sets of $\mathbb{R}$, a topology for $P^1(R)$ is obtained, and then there will be no difficulty in proving that the stereographic projection illustrated in Figure 3.3 is a homeomorphism.

5.6 Some results on the circle

In this appendix we can see the enormous advantage of working with complex coordinates.

Inversion with respect to a Euclidean circle

The conjugate of a complex number $z = x + iy$ is the complex number in which the real part remains the same and the imaginary part changes sign, $\bar{z} = x - iy$. The geometric interpretation is a reflection with respect to the real axis.

Consider now a Möbius transformation T that maps $\mathbb{R}$ into S^1, such as

$$T(z) = \frac{z+i}{iz+1}.$$

It is clear that if z is in $\mathbb{C} - \mathbb{R}$, then the points $T(z)$ and $T(\bar{z})$ are such that one is in the "bounded side" of S^1 and the other is outside this region. In fact we can verify that the product of their norms is 1:

$$\left|\frac{z+i}{iz+1}\right|\left|\frac{\bar{z}+i}{i\bar{z}+1}\right| = \sqrt{\left(\frac{z+i}{iz+1}\right)\left(\frac{\bar{z}-i}{-i\bar{z}+1}\right)\left(\frac{\bar{z}+i}{i\bar{z}+1}\right)\left(\frac{z-i}{-iz+1}\right)}$$

$$= \sqrt{\left(\frac{z^2+1}{z^2+1}\right)\left(\frac{\bar{z}^2+1}{\bar{z}^2+1}\right)} = 1; \tag{5.1}$$

the second equality above comes from the fact that

$$(z+i)(z-i) = (iz+1)(-iz+1) = z^2 + 1$$

and similarly for $\bar{z}$.

In plane Euclidean geometry, given a circle $\mathcal{C}$ with center O and radius R, we call two points P and P', such that

$$d(O,P)d(O,P') = R^2,$$

inverse points with respect to $\mathcal{C}$.

Equation (5.1) shows that $T(z)$ and $T(\bar{z})$ are inverse points with respect to S^1, and our experience with the Möbius transformation shows us that if after applying the inversion with respect to S^1 we apply a homothesis H of ratio λ, the product of the norms of $H \circ T(z)$ and $H \circ T(\bar{z})$ would have been λ^2.

That is why it makes sense to call the inversion with respect to a Euclidean circle a **reflection with respect to a hyperbolic line**. That is, inversions on circles play the same role in hyperbolic geometry as the role played in Euclidean geometry by reflections on straight lines.

It is interesting to note that the points P' and P'' corresponding to the stereographic projections of a point $P \in S^2$ from the north pole, N, and the south pole, S, respectively, are inverse points with respect to the circle of radius 1 of the equator. This is easy to prove if we consider the right triangles that appear in the figure created on the plane containing all the points we are dealing with: the center O of S^2, N, S, P, P' and P''.

We invite the reader to make the drawing and to prove the previous result, as well as to consider that being P' and P'' inverse points with respect to S^1 proves that S^2 is a differentiable manifold, for the change of coordinates from P' to P'' is given by $z \mapsto 1/z$: a diffeomorphism if $z \neq 0$.

Circles of Apollonius and Steiner networks

A **circle of Apollonius** is the locus of the points P of a plane whose distances to two fixed points, α and β, are equal to a constant k.

In complex coordinates we write

$$\frac{|z-\alpha|}{|z-\beta|} = k, \qquad (5.2)$$

Equation (5.2) makes sense for every $k \in \mathbb{R}^+ \cup \{0\}$, and written in Cartesian coordinates, it takes the form

$$(1-k^2)x^2 + (1-k^2)y^2 - 2(\alpha + \beta k^2)x + \alpha^2 - k^2\beta^2 = 0, \qquad (5.3)$$

which is the equation of a circle because it does not have an xy term and the coefficients of x^2 and y^2 are equal.

As k varies in $\mathbb{R}^+ \cup \{0\}$, we obtain the family of circles of Apollonius determined by the points α and β (it is the family $\mathcal{F}_2$ in Figures 4.9 and 4.10 (a)).

From equation (5.2) it is immediate that as $k \to 0$, the circle approaches the point α, and that as $k \to \infty$, the circle approaches the point β. These points are called the **limit points** of the family of circles of Apollonius.

Also, when $k > 1$, β is in the interior of the circle, although β is not its center, and when $k < 1$, α is in the interior of the circle, again, although α is not its center.

Equation (5.3) shows that when $k = 1$, the circle is a straight line, the line bisector of the segment defined by α and β.

The transformation $T \in PSL(2, \mathbb{C})$

$$T(z) = \frac{z-\alpha}{z-\beta},$$

maps α into 0 and β into ∞, and accordingly transforms the circles through those points (the family $\mathcal{F}_1$ in Figures 4.9 and 4.10) into Euclidean straight lines through the origin, whereas the circles of Apollonius (the elements of the family $\mathcal{F}_2$) are transformed into concentric circles with its center at the origin, for if $w = T(z)$ is such that $|w| = \rho$, that is equivalent to

$$\frac{|z-\alpha|}{|z-\beta|} = \rho;$$

that is, z belongs to the circle of Apollonius corresponding to the constant ρ.

The families $\mathcal{F}_1$ and $\mathcal{F}_2$ form a **Steiner network**, and its elements are called **Steiner circles**. The properties listed in [A] for these circles (and which follow from what we have discussed so far) are the following:

1. For each point in $\widehat{\mathbb{C}} - \{\alpha, \beta\}$, there is one and only one element $C_1 \in \mathcal{F}_1$ and only one element $C_2 \in \mathcal{F}_2$ which contain it.

2. Any $C_1 \in \mathcal{F}_1$ and any $C_2 \in \mathcal{F}_2$ intersect at right angles.

3. The Euclidean inversion (hyperbolic reflection) in $C_1 \in \mathcal{F}_1$ maps any $C_2 \in \mathcal{F}_2$ into itself, and to each $C_1 \in \mathcal{F}_1$ into another element of the same family.

4. The limit points are symmetric (inverse to one another) with respect to each element $C_2 \in \mathcal{F}_2$, and only with respect to them.

Bibliography

- **A** Ahlfors, L.V. *Complex Analysis*, Mc Graw Hill, 1985.
- **Be** Beardon, A.F. *The Geometry of Discrete Groups*, Springer-Verlag, 1983.
- **B-ML** Birkhoff, G., MacLane, S. *A Survey of Modern Algebra*, MacMillan, 1963.
- **Bo** Bonola, R. *Non-Euclidean Geometry*, Dover, 1955.
- **C-G** Costa, A., Gómez, B. *Arabesques and Geometry*, Springer VideoMATH, 1991.
- **Cou** Courant, R. *Differential and Integral Calculus*, Interscience, 1937.
- **Cox1** Coxeter, H.S.M. *Fundamentos de Geometría*, Limusa, 1971.
- **Cox2** Coxeter, H.S.M. *Regular Complex Polytopes*, Cambridge University Press, 1974.
- **Cox3** Coxeter, H.S.M. *Projective Geometry*, Springer-Verlag, 1994.
- **Cox4** Coxeter, H.S.M. *Non-euclidean Geometry*, University of Toronto Press, 1965.
- **Cox5** Coxeter, H.S.M. *Geometry Revisited*, The Mathematical Association of America, 1983.
- **DoC** Do Carmo, M.P. *Differential Geometry of Curves and Surfaces in $\mathbb{R}^3$*, Prentice Hall, 1976.
- **D-P** Dodson. C., Parker, E. *A User's Guide to Algebraic Topology*, Kluwer Ac. Pub. Group, 1997.
- **Du** Durero, A. *Instituciones de Geometría*, Fuentes **3**, IIB, UNAM, 1987.
- **Er** Ernst, B. *El Espejo Mágico de M.C. Escher*, Taschen, 1992.
- **Es** Escher, M.C. *The Graphic Work of M.C. Escher*, Taschen, 1992.
- **Eu** Euclid. *Euclid's Elements*, Dover, 1979.
- **Ev** Eves, H. *Estudio de las Geometrías*, UTEHA, 1969.
- **F** Forster, O. *Lectures on Riemann Surfaces*, Springer-Verlag, 1991.

- **Fu** Fulton, W. *Algebraic Curves*, Mathematics Lecture Notes Series, Benjamin, 1969.
- **G-R** González, V., Rodríguez, R. *Seminario de Geometría Compleja 1*, Universidad de Santa María y Universidad de Santiago, 1983.
- **G-S** Grünbaum, B., Shepard, G.C. *Tillings and Patterns*, W.H.Freeman and Co., 1987.
- **Gr** Graustein, W. *Introduction to Higher Geometry*, The Macmillan Co., 1949.
- **H** Hilbert, D. *Foundations of Geometry*, Open Court Publishing Co., 1962.
- **H-C** Hilbert, D., Cohn Vossen, S. *Geometry and the Imagination*, Chelsea Publishing Co., 1983. (Edición facsimilar en Vínculos Matemáticos No. 150, FC, UNAM, 2000.).
- **Hr** Hirsch, M. *Differential Topology*, Springer-Verlag, 1967.
- **I** Illanes, A. *La Caprichosa Forma de Globión*, la ciencia para todos, **168**, Fondo de Cultura Económica, 1999.
- **Ke** Klein, F. *Le Programme d'Erlangen*, Gauthier-Villards, Paris, 1974.
- **Ki** Kline, M. *Mathematical Thought from Ancient to Modern Times*, Oxford University Press, 1991.
- **L** Lefschetz, S. *Differential Equations: Geometric Theory*, Pure and Applied Mathematics **VI**, Interscience, 1957.
- **Ma** Markushevich, A. *Teoría de las Funciones Analíticas*, MIR, 1987.
- **Mar** Martin, G. *Transformation Geometry. An Introduction to Symmetry*, Springer-Verlag, 1997.
- **Mat** Matsushima, Y. *Differential Manifolds*, Marcel Dekker, 1972.
- **Mo** Montesinos, J.M. *Classical Tessellations and three-Manifolds*, Springer-Verlag, 1985.
- **Ra** Ramírez-Galarza, A. *Geometría Analítica: Una introducción a la Geometría*, Las Prensas de Ciencias, UNAM, 1998.
- **R-S** Ramírez-Galarza, A., Sienra, G. *Invitación a las Geometrías No-euclidianas*, Las Prensas de Ciencias, UNAM, 2000.
- **Re** Rees, E. *Notes on Geometry*, Universitexts, Springer Verlag, 1983.
- **Ri** Rincón, H. *Álgebra Lineal*, Las Prensas de Ciencias, UNAM, 2001.
- **Sa** Samuel, P. *Projective Geometry*, Springer-Verlag, 1988.
- **Se** Seade, J. *On the topology of isolated singularities in analytic spaces*, Birkhäuser, Progress in Mathematics vol. 241, 2006.

- **Sp** Speiser, A. *Die Theorie der Gruppen von Endlicher Ordnung*, Dover, 1945.
- **Spr** Springer, G. *Introduction to Riemann Surfaces*, Addison-Wesley Publishing, 1957.
- **T** Thurston, W. *Three Dimensional Geometry and Topology*, Vol. 1, Princeton Mathematical Series **35**, 1997.
- **Va** Vasari, G. *Vidas de los más excelentes pintores, escultores y arquitectos*, Nuestros Clásicos **74**, UNAM, 1996.
- **Ve** Verjovsky, A. *Introducción a la Geometría y las Variedades Hiperbólicas*, Departamento de Matemáticas, CINVESTAV, IPN, 1982.
- **W** Wolfe, H.E. *Introduction to Non-euclidean Geometry*, Holt, Reinhart and Wiston, 1945.
- **Y** Yaglom, I. *Felix Klein and Sophus Lie, evolution of the idea of symmetry in the nineteenth century*, Birkhäuser, 1988

Index

action of a group, 45
algebraically closed fields, 127
alternate group, 204
analytic function, 188
angle
    of parallelism, 180
angle
    between two vectors, 7
antipodal, 25, 96
arc length, 34
area of a parallelogram, 7
asymptote, 87
attractor point, 167
automorphism, 159
axis
    of a rotation, 25
    of symmetry, 3

base
    standard, 7
basis
    left, 25
Beltrami, E., 150
biholomorphism, 112
Bolyai, J., 149
boundary of a set, 39
bounded set, 207

$\mathbb{C}$-differentiable function, 188
Cartesian plane, 77
center
    n-, 67
    of a cube, 13
    of symmetry, 3
change of coordinates, 110

circle
    maximal, 141
    of Apollonius, 165, 209
class
    of lines, 80
    of pairs, 80
cm, 72
cmm, 70
complex plane
    extended, 95
conformal, 96, 98
conformally equivalent, 190
conjugate
    subgroup, 65
conjugation, 65, 159
coordinate chart, 110
covering
    double, 96
critical point, 200
critical value, 33, 200
cross ratio, 121, 123
    of four points in a conic, 131
crossing
    straight lines, 15
curvature
    constant, 36
    maximal, 35
    minimal, 35
    of a curve, 34
    of a surface, 32, 35
    sectional, 35
cycle, 202

dense set, 43

derivative
 at P_0, 199
 partial, 199
diffeomorphism, 110
differentiable manifold, 110
dihedral group, 48
discontinuous subgroup, 183
distance
 from a point to a plane, 2
 from a point to a straight line, 2, 7
 from a point to another, 2, 6
 of translation, 185
 oriented, 123
domain
 fundamental, 62
dual
 of a platonic solid, 50
 space of a vector space, 100
dual proposition, 99
duality
 principle, 99

edges, 49
elliptic point, 34
elliptic transformation, 166
epicycloid, 44
equivalence
 class, 37
equivalence relation, 201
Escher, M.C., 105
Euclid, 1
Euler
 characteristic of, 52
Euler, L., 52

face, 49
fibration of Hopf, 98
foliation, 43, 200
frieze patterns, 58
function
 analytic, 112
 continuous, 207
 holomorphic, 112

fundamental region, 185

Galois, E., 15
Gauss, K.F., 127
Gaussian curvature, 35
general position, 98
generators of a group, 47
genus, 52, 189
geodesic, 39, 141
gradient, 199
Grassmannian manifold, 103
group, 18
 Abelian, 18
 action of a, 45
 affine, 84
 commutative, 18
 cyclic, 48
 Euclidean, 26
 Lie, 24
 linear, 113
 of direct isometries, 159
 of symmetries, 47
 orthogonal, 23, 26
  special, 23
 projective, 113
 relations in a, 47
 transitive, 158

handle, 188
harmonic
 conjugates, 122
 set, 122
Hilbert, D., 150
holomorphic function, 112, 188
homeomorphism, 94, 207
homogenizing, 111
homothesis, 28
horizon line, 75, 77
horocycle, 167
horosphere, 181
hyperbolic
 line, 151
 constant, 179
 length, 170

metric, 170
norm, 169
parallels, 153
point, 34
transformation, 166
hypersurface, 128
hypocycloid, 44

ideal polygons, 156
image,
    inverse, 200
incidence, 85, 99
inclusion
    standard, 7
injective
    transformation, 17
interior
    of a set, 39
inverse
    image, 33
inverse points, 208
inversion, 208
involution, 64
isometry, 16
    direct, 159
isotropy, 36
iteration, 46

Klein bottle, 109
Klein, F., 16

Law of Refraction, 153
leaf, 200
length
    of a curve, 141
level set, 201
level surfaces, 32
Lie group, 18
line
    elliptic, 142
Lobachevski, N., 149

Möbius
    strip, 104

Möbius, F. A., 104
manifold, 37, 184
matrix
    of the conic, 126
    of the metric, 134
    of the quadratic form, 29
    orthogonal, 23
    rank of a, 29
    signature of a, 29
    transposed, 23
metric, 134
model
    complete, 150
    of Beltrami, 152
    of the Poincaré disk, 152
    of the hyperboloid, 152
    of the upper half plane, 153
    projective, 151

n-cycle, 203
neighborhood, 207
network of Steiner, 165
norm
    of a vector, 6
normal
    subgroup, 46
normal section, 34

open set, 206
open,
    relative, 207
orbit, 45

p1, 72
p2, 70
p3, 68
p31m, 69
p3m1, 69
p4, 69
p4g, 70
p4m, 69
p6, 68
p6m, 68
parabolic transformation, 166

parametrization, 110
Pascal, B., 129
pencil
    of lines, 101
permutation, 202
perspectivity, 113
pg, 72
pgg, 70
plane
    affine, 79
    elliptic, 141
    of symmetry, 4
    of the drawing, 77
    real projective, 93
platonic solid, 49
Playfair,
    Axiom of, 205
pm, 72
pmg, 70
pmm, 72
Poincaré, H., 152
point
    at infinity, 79
    navel or umbilic, 36
    vanishing, 77
point(s)
    at infinity, 151, 154
    interior, 38
    parabolic, 34
points
    limit, 209
polar
    line, 135
    point, 135
polyhedron
    regular, 49
product
    cross, 7
    scalar, 6
    scalar triple, 7
projection
    standard, 41
    stereographic, 95

projective
    line, 97
    point, 93, 94
projectivity, 113
projectivized space, 93
pseudosphere, 150

quadric
    degenerated, 32
    surface, 29
quadrilateral, 75
    vertex of a, 75
quotient set, 201
quotient space, 184

range
    of points, 101
rank, 29
ratio of division, 88
real projective
    line, 94
reflection
    in $\mathbb{R}^3$, 25
    in a hyperbolic line, 208
    with respect to a line, 22
region
    fundamental, 38
regular value, 33
    inverse image of a, 33
repelling point, 167
residual class, 201
Riemann
    sphere, 95
Riemannian metrics, 170
rotation, 19, 24
    axis of a, 25

Schwarz inequality, 7
separatrix, 201
signature, 29
space
    of identification, 45
spherical geometry, 142
spindle, 144

stabilizer, 58
standard form
    of a hyperbolic transformation, 166
    of a parabolic transformation, 166
Steiner
    circles, 210
    network, 165, 210
    theorem of, 130
Steiner network
    degenerated, 165
step, 61
straight line
    at infinity, 79
subgroup, 45
    alternate, 57
    stabilizer, 58
surface
    extension of a, 30
    homogeneous, 36
    of constant curvature, 36
    of Veronese, 128
    orientable, 106
    Riemann, 112
    ruled, 30
symmetric group, 53, 202
symmetry with respect to
    a coordinate axis, 5
    a plane, 4
    a point, 3
    a straight line, 3
    the origin, 5

tangent
    line, 136
    plane, 33
tessellation, 59
Theorem
    (of the) Mystic Hexagon (Pascal), 131
    (Fundamental) of Projective Geometry, 117
    of Bezout, 139
    of classification of surfaces, 188
    of Desargues, 101
    of Pappus, 129
    of Steiner, 130
    of the Crystallographic Restriction, 67
    of uniformization, 189
    Riemann's mapping, 190
Thurston, W., 149
tiling, 59
topological space, 207
topology, 207
torus, 183, 189
    of revolution, 33
    plane, 40
tractrix, 150
transformation
    affine, 83
    identity, 17
    Möbius, 120
    rigid, 16
transitivity, 114
translation, 16, 24
transposition, 202
triangle, 99
    self-polar, 137
trilateral, 99
triple scalar product, 7

ultraparallels, 153

value,
    regular, 200
vanishing points, 75
Veronese mapping, 128
vertex, 49